GREAT FACTS:

A

POPULAR HISTORY AND DESCRIPTION

OF THE MOST

REMARKABLE INVENTIONS

DURING THE PRESENT CENTURY.

BY

FREDERICK C. BAKEWELL,

AUTHOR OF

"PHILOSOPHICAL CONVERSATIONS," "MANUAL OF ELECTRICITY," ETC.

ILLUSTRATED WITH NUMEROUS ENGRAVINGS.

NEW YORK:

D. APPLETON AND COMPANY,

346 & 348 BROADWAY.

1860.

GREAT FACTS.

PREFACE.

The conveniences, the comforts, and luxuries conferred on Society by the many important Inventions of the present century, must naturally excite a desire to know the origin and progress of the application of scientific principles, by which such advantages have been gained.

Practically considered, those Inventions are of much greater value than the discoveries of Science on which most of them depend; and the scientific inquirer who confines his views to abstract principles, without looking beyond them to the varied methods of their application to useful purposes, may be compared to a traveller who, having toiled arduously to gain the top of a mountain, then shuts his eyes on the prospect that lies before him.

To the inquiring youth, more particularly, it is desirable that he should be enabled to satisfy his wish to know by what means such wonders as Steam Navigation, Locomotion on Railways, the Electric Telegraph, and Photography have been gradually developed; and in becoming acquainted with the successive steps by which they have advanced towards their present perfection, he will at the same time learn a useful lesson of perseverance under difficulties, and will have his mind impressed with many valuable scientific truths. The knowledge to be gained by such inquiry is eminently practical, and of a kind which those engaged in any of the pursuits of life can scarcely fail to require.

A History of Inventions almost necessarily implies a description of the mechanisms and processes by which they are effected; so far, at least, as to render the principles on which their actions depend understood. It would be impossible, however, in a work of this limited size to enter minutely into explanations of mechanisms, and into the applications of scientific discoveries, which would require a separate treatise for each; but it has been the Author's endeavour to give a succinct, intelligible account, free from technicalities, of the manner in which they operate, so as to be comprehensible to all classes of readers.

By thus giving a popular character to the work, to make it acceptable to the young, it is hoped that it will not be found less worthy, on that account, the perusal of those more advanced in life.

When Beckman wrote his History of Inventions, towards the close of last century, scarcely any of the wonderful discoveries and contrivances that now form parts of our social system were known; and the table of contents of his two large volumes affords a curious insight to the nature and limited extent of such contrivances as were then considered most important. The introduction into his history of such subjects as canary birds, carp, the adulteration of wine, apothecaries, cock-fighting, and juggling, lead us to infer that the Historian of Inventions at that time must have had some difficulty to find appropriate matter wherewith to fill his volumes. The opposite difficulty now presents itself. The numerous important, wonderful, and curious accomplishments of human skill and ingenuity during the present century render preference perplexing, where so many deserve description. From among the number that press for notice, the Author has endeavoured to select those that are either the most important, the most remarkable, or that seem to possess the germs of future progress;

and he trusts that the selection he has made, and the mode in which the subjects have been treated, will render this volume interesting and instructive.

F. C. B.

6 *Haverstock Terrace, Hampstead,*
November, 1858.

CONTENTS.

GREAT FACTS.

THE PROGRESS OF INVENTION.

THE inventive faculty of man tends more directly than any other intellectual power he possesses to raise him in the scale of creation above the brutes. Nearly every advance he makes beyond the exercise of his natural instincts is caused by invention—by that power of the mind which combines known properties in different ways to obtain new results.

When an Indian clothes himself with the skins of animals, and when he collects the dried leaves of the forest for his bed, he is either an original inventor, or he is profiting by the inventions of others. Those simple contrivances—the first steps in the progress of invention—are succeeded by the more labored efforts of inventive genius, such as contriving means of shelter from rain, or from the heat of the sun, when caves cannot be found to creep into, or the overhanging foliage fails to afford sufficient covering. The construction of places of shelter is an imitation of the protection formed by Nature; and the rudest hut and the most magnificent palaces have their prototypes in caverns and in the interlacing branches of trees.

Nature also supplies knowledge of the means by which inventors are enabled to work. The savage who seizes hold of a broken bough is in possession of the *lever*, the uses of which he learns by the facility it affords in moving other objects. He ascends to the top of a precipice by walking up the sloping hill behind, and he thus becomes practically acquainted with the principle of the *inclined plane*. The elements of all the mechanical powers are then at his command, to be applied by degrees in administering to his wants, as his inventive faculties, guided by observation and experience, suggest. An accidental kick against a loose stone shows the action of propulsive force; and the stone that he has struck with his foot, he learns to throw with his hand. The bending of the boughs of trees to and fro by the wind teaches the action of springs; and in the course of time the bow is bent by a strip of hide, and the relaxation ot the spring, after farther bending, propels the arrow. Observation and imitation thus lead to invention, and every new invention forms the foundation of further progress.

It has been so with every invention at present known, and must so continue to the end of time:—"There is nothing new under the sun." Gas lighting, Steam locomotion, and the Electric Telegraph have each sprung from some source "old as the hills," though so modified by gradually progressive changes, that the giant we now see bears no resemblance to the infant of ages past.

The observation that light particles floating in the air are attracted by amber when rubbed, which was made known six centuries before the Christian era, was the origin of the invention by which communications are now transmitted, with the rapidity of lightning, from one part of the world to another. There is no apparent relation between effects so dissimilar; yet the steps of progress can be distinctly traced, from the attraction of a feather to the development of the electric telegraph.

Whenever the history of an invention can be thus tracked backward to its source, it will be found to have advanced to its present state by progressive steps, each additional advance having been dependent on the help given by the progress before made. Sometimes these onward movements are greater and more remarkable than others, and the persons who made them have become distinguished for their inventive genius, and are considered the benefactors of mankind; yet they were but the followers of those who had gone before and shown the way.

Many of the most remarkable inventions are attributable to accidents noted by observing and inventive minds. Not unfrequently also have important discoveries of truth been made in endeavouring to establish error; and new light is being constantly thrown on the path of invention by unsuccessful experiments.

This view of the means by which inventions

originate and are brought to perfection may appear to detract from the merit of inventors, since it regards them as founding their conceptions altogether on the works of others, or on chance. But instead of diminishing their claims to approbation and reward, it places those claims on a more substantial foundation than that of abstract original ideas. The man who has the faculty to perceive that by a different application of well-known principles he can produce useful effects before unknown, directly benefits mankind far more than the discoverer of the principles which had till then lain dormant; and the numerous difficulties which ever arise before an invention can be practically operative, frequently afford exercise for reasoning powers of the highest kind, which may develop new arrangements, that exhibit as much originality and research as were displayed by the discoverers of the principles on which the invention depends.

The dependence of every invention on preceding ones produces very frequently conflicting claims among inventors, who, forgetting how much they were indebted to others, do not hesitate to charge those, who make still further improvements, with imitation and piracy. It is, indeed, sometimes difficult to determine whether the alterations made in well-known contrivances are, or are not, of sufficient importance to constitute inventions; and there can be no doubt that there is too great facility afforded, by the indiscriminate grant of letters patent, for the establishment of monopolies that often serve to obstruct further improvements. At the same time, it must be observed that a very trifling addition or change occasionally gives practical value to an invention, which had been useless without it. In such cases, though the individual merit of the inventor is small, the benefit conferred may be important, and may operate influentially in promoting the progress of civilization.

Scientific discovery goes hand in hand with invention, and they mutually assist each other's progress. Every discovery in science may be applicable to some new purpose, or give greater efficiency to what is old. Those new and improved instruments and processes provide science with the means of extending its researches into other fields of discovery; and thus, as every truth revealed, supplies inventive genius with fresh matter to mould into new forms, those creations become in their turn agents in promoting further discoveries.

The action and reaction thus constantly at work, tend to give accelerating impulse to invention, and are continually enlarging its sphere of operations. Instead, therefore, of supposing, as some do, that invention and discovery have nearly reached their limits, there is more reason to infer that they are only at the commencement of their careers; and that, great as have been the wonders accomplished by the applications of science during the first half of the present century, they will be at least equalled, if not surpassed, by those to be achieved before its close.

STEAM NAVIGATION.

SHIPS, propelled by some mysterious power against wind and against tide, cutting their ways through the water without apparent impulse and like things of life, were not unfrequently seen gliding along in the regions of fancy, ages before the realization of such objects on geographical seas and rivers was looked upon as in the slightest degree possible. Even at the beginning of the present century, it seemed to be more probable that man would be able to navigate the air at will, than that he should be able, without wind or current, and in opposition to both, to propel and steer large ships over the waves; yet, within twenty years afterwards, Steam Navigation had ceased to be a wonder.

If we look back into the records of past ages, we find that inventive genius was active in the earliest times, in endeavouring to find other means of propelling boats than by manual labour and the uncertain wind, some of which contrivances point to the method subsequently adopted by the constructors of steam-vessels.

To enable us to appreciate properly the gradual advances that have been made in perfecting any invention, it is necessary to consider its distinguishing features, and the difficulties which inventors have had successively to contend against. On taking this view of the progress of Steam Navigation, it will be found that the amount of novelty to which each inventor has a claim is very small, and that his principal merit consists in the application of other inventions to accomplish his special object. The same remark will indeed apply to most other inventions; for the utmost that inventive genius can accomplish, is to put together in new forms, and with different applications, preceding contrivances and discoveries, which were also the results of antecedent knowledge, labour, and skill.

When, for instance, we look upon an ordinary steamboat, the most remarkable and the most important feature is the paddle-wheel, by the action of which against the water the boat is propelled. Yet that method of propelling boats was practised by the Egyptians hundreds of years before steam power was thought of; and the ancient Romans made use of similar wheels, worked by hand, as substitutes for oars. It would seem, therefore, to be only a small step in inventive progress, after the discovery of the steam engine, to apply that motive power to turn the paddle-wheels which had been previously used; and now that we see the perfected invention, it may surprise those who are unacquainted with the difficulties which attend any new appliance, that Steam Navigation did not sooner become an accomplished fact.

In a book called "Inventions and Devices," by William Bourne, published in 1578, it was proposed to make a boat go by paddle-wheels, "to be turned by some provision." The Marquis of Worcester, in his "Century of Inventions," also speaks vaguely of a mode of propelling ships. But Capt. Savery, the inventor of the earliest working steam engine, was the first to suggest the application of steam to navigation; and Dr. Papin, who contended with Savery for priority of the invention, also suggested about the same time the application of the elastic force of steam to that purpose.

These crude notions, however, do not deserve to be considered as inventions, though they probably assisted in suggesting the idea of the plan proposed by Mr. Jonathan Hulls, who in 1736 took out a patent for a steam-boat, and in the following year published a description of his invention, illustrated by a drawing, entitled, "A description and draught of a new-invented machine for carrying vessels or ships out of or into any harbour, port, or river, against wind or tide, or in a calm."

The greater part of this publication is occupied with answers to objections that he supposed might be raised to the scheme, and in the preface he makes the following observations on the treatment inventors were exposed to in his day, which we fear will apply equally at the present time. "There is," he says, "one great hardship lies too commonly on those who purpose to advance some new though useful scheme for the public benefit. The world abounding more in rash censure than in candid and unprejudiced estimation of things, if a person does not answer their expectations in every point, instead of friendly treatment for his good intentions, he too often meets with ridicule and contempt."

At the time of Mr. Hulls' invention, Watt had not made his improvements in the steam engine, and the kind of engine Hulls employed was similar to Newcomen's, in which the steam was condensed in the cylinder, and the piston, after being forced down by the direct pressure of the atmosphere, was drawn upwards again by a weight. The paddle, or "vanes,"

as he called them, were placed at the stern, between two wheels, which were turned by ropes passing over their peripheries. The alternate motion of the piston was ingeniously converted into a continuous rotary movement, by connection with other ropes attached to the piston and to the weight, the backward movement being prevented by a catch or click.

The woodcut which lays before you is a reduced

copy of Hulls' "draught" of his steam-boat, as given in his book, a copy of which is preserved in the British Museum.

The utmost application of steam power to navigation contemplated by Hulls was to tow large vessels into or out of harbour, in calm weather, by means of a separate steam tug-boat, as he considered the cumbersome mechanism would be found objectionable on board the ships to be thus propelled. It does not appear that this plan was effectually tried, nor was the arrangement of the mechanism, nor the imperfect condition of the steam engine at that period, calculated to make the effort successful.

For some years after Mr. Hulls' plan had been published, and had proved abortive, no further attempt seems to have been made, until the improvements in the steam engine, by Watt, rendered it more applicable for the purpose of navigation. The French claim for the Marquis de Jouffroy the honour of having been the first who successfully applied steam power to propel boats, in 1782; though another French nobleman, the Comte d'Auxiron, and M. Perier, had eight years previously made some experiments with steam-boats on the Seine. The Marquis de Jouffroy's steam-boat, which was 145 feet long, was tried on the Soane, near Lyons, with good promise of success. The marquis was, however, obliged to leave France by the fury of the Revolution, and when he returned in 1796, he found that a patent had been granted to M. le Blanc, for building steam-boats in France. He protested against the monopoly, but the patent remained in force, and the plan received no further development, either from the Marquis de Jouffroy, or the patentee.

About five years later, Mr. Patrick Miller, of Dalswinton, in Scotland, directed his attention to the propulsion of boats by mechanical means, and contrived different kinds of paddles, and other propellers to be worked by hand, which were tried on boats on Dalswinton Lake. The great labour required to work these machines induced Mr. James Taylor, a tutor in Mr. Miller's family, to suggest the use of steam power to turn them, and he recommended Mr. Miller to obtain the assistance of William Symington, an engineer, who was at that time endeavouring to make a steam locomotive carriage. Among the first difficulties that suggested themselves, was the danger of setting fire to the boat by the engine furnace. This difficulty was overcome by Mr. Taylor, and the arrangements were completed, and the experiment was tried in 1788 The steam engine and mechanism were applied to a double pleasure-boat; the engine being placed on one side, the boiler on the other, and the paddle-wheel in the centre. The cylinders of the steam engine were only four inches in diameter; but with this engine the boat was propelled across Dalswinton Lake at a speed of five miles an hour.

The success of this experiment induced Mr. Miller to have a larger boat built, expressly adapted for the introduction of a steam engine. It was constructed under the superintendence of Symington, and was tried successfully on the Forth and Clyde Canal in

1789, when it was propelled at the rate of seven miles an hour.

In the arrangement of the mechanism of this boat, the cylinder was placed horizontally, for the purpose of making connection between the paddle-wheel and the piston, without the working beam. The piston was supported in its position by friction wheels, and communicated motion to the paddles by a crank. The paddles were placed in the middle of the boat, near the stern; and there was a double rudder, connected together by rods which were moved by a winch at the head of the vessel.

It is not very clear why Mr. Miller did not follow up this success. Objection, indeed, was made by the proprietors of the canal on account of the agitation of the water, which it was feared would injure the banks. It would appear also that a misunderstanding took place between Miller and Symington, which gave the former a distaste to the undertaking; and having shown that such a plan was practicable, he left others to carry it into practical effect.

Several methods of propelling boats, otherwise than by paddles, had some years previously been suggested; among which were two that have been again and again tried by succeeding inventors, down to the present day.

One of these is an imitation of the duck's foot, which expands when it strikes the water, and collapses when it is withdrawn. The other is the ejection of a stream of water at the stern, or on both sides of the boat, so as to produce a forward movement by reaction. Both these plans of propulsion seem feasible in design; but they have hitherto failed in practice. A pastor at Berne, named J. A. Genevois, has the credit of having invented the duck-feet propeller in 1755; and in 1795, six years after Mr. Miller's successful experiments, Earl Stanhope had a steam-boat built on that principle. It was so far a failure, that it was not propelled faster than three miles an hour. The other method of propulsion, though of older date, was patented in 1800 by Mr. Linnaker, who proposed to draw the water in at the head of the vessel, and eject it at the stern, and thus to obtain a double action on the water for propelling; but the plan was not found to answer.

In 1801, Lord Dundas revived Mr. Miller's project, and availed himself of Mr. Symington's increased experience and the further improvements in the steam engine, to construct a much more perfect steam-boat than any that had been made. He spent £3,000 in the experiments, and in March, 1802, his vessel, called the "Charlotte Dundas," was tried on the same scene of action, the Forth and Clyde Canal. This boat, according to Symington's report, towed two vessels, each of seventy tons burthen, a distance of nineteen miles and a half in six hours, against a strong wind. The threatened injury to the banks of the canal by the great agitation of the water prevented the use of this boat, which was consequently laid aside; for the views of the inventors of steam-boats in the first instance were limited to their employment to drag boats along canals.

We now approach a period when more decided advances and more rapid progress were made towards realizing steam navigation as a practical fact. Mr. Fulton, an American, residing in France, after making a number of experiments, under the sanction and with the assistance of Mr. Livingstone, the American Ambassador, launched a small steam-boat on the Seine in 1803, but the weight of the engine proved too great for the strength of the boat, which broke in the middle, and immediately went to the bottom.

Not disheartened by this failure he built another one, longer and stronger, and this he succeeded in propelling by steam power, though very slowly. It was, indeed, a much less successful effort than the attempts of Mr. Miller and Lord Dundas. Having been threatened with opposition by M. le Blanc, the patentee of steam-boats in France, Fulton determined to return to his native country, where the large navigable rivers and lakes offered ample scope for the development of steam navigation. Having heard of the success of Symington's boats, he visited Scotland for the purpose of profiting by his experience; and he induced Symington, by promises of great advantages if the invention succeeded in America, to show him the "Charlotte Dundas" at work, and to enter into full explanations of every part. Thus primed with the facts, and with the further suggestions of Symington, Fulton repaired to New York. Mr. Livingstone, who had assisted Fulton in his experiments, was himself an inventor of several plans of propelling vessels by steam, and in 1798 he obtained a patent in the State of New York, for twenty years, on condition that he should produce a steam-boat by the 7th of March, 1799, that would go at the rate of *four* miles an hour. Having failed to fulfil that condition, the patent privilege was left open, and was promised to the first inventor who succeeded in propelling a boat by steam power at the proposed speed of four miles an hour. Fulton, who had entered into partnership with Mr. Livingstone, possessed advantages in the construction of the vessel he built in America, far greater than any previous inventor. He had not only gained knowledge by his former failures, but he was able to profit by the experience of others, and he had secured a superior steam-engine, manufactured by Boulton and Watt, of twenty-horse power. This was a much more powerful engine than any that had been used in any former experiment; the one employed by Mr. Livingstone having had only five-horse power. This steam-vessel was launched at New York in 1807, and was called the "Clermont," the name of Mr. Livingstone's residence on the banks of the Hudson. Its length was 133 feet, depth 7 feet, and breadth 18 feet. The boiler was 20 feet long, 7 feet deep, and 8 feet broad. There was only one steam cylinder, which was 2 feet in diameter, with a length of stroke of 4 feet. The paddle-wheels were 15 feet in diameter, and 5 feet broad; and the burthen of the vessel was 160 tons. Crowds of spectators assembled to see the boat start on its first experimental voyage. The general impression, even of those who were friendly to Fulton, was that it would fail, and an accident which occurred when the vessel was under way confirmed this opinion. The foreboders of evil exclaimed immediately that they had "foreseen something of the kind;" and observed "it was a pity so much expense had been incurred for nothing!" The required repairs were, however, soon made. The vessel when again tried cut her way bravely through the water, to the astonishment of all, and the doubts, and fears, and lamentations were quickly changed into congratulations.

As the "Clermont" urged its way up the Hudson, its chimney emitting innumerable sparks from the dried pine wood used as fuel, it excited great alarm among those who were not prepared for such an apparition. An American paper of that day thus described the effect produced on the crews of other ships in the river:—"Notwithstanding the wind and tide were adverse to its approach, they saw with astonishment that it was rapidly coming towards them; and when it came so near that the noise of the machinery and paddles was heard, the crews, in some instances, shrunk beneath their decks from the terrific sight, or left their vessels to go on shore; whilst others prostrated themselves and besought Providence to protect them from the approach of the horrible monster which was marching on the waves, and lighting its path by the fires which it vomited."

During the time that Fulton was building his steam-boat Mr. R. L. Stevens, of Hoboken, in the State of New Jersey, was also engaged in a similar undertaking. Though his name is comparatively little heard of in the history of Steam Navigation, his efforts were more successful than any that had been made previously, and but for the fortunate chance to Fulton that he was able to launch and put his boat in action a few days before Stevens had completed his, all, and more than all, the merit that is now ascribed to the former would have been attributed to Stevens. The previous successful experiment of Fulton having fulfilled the conditions imposed by the State of New York, he obtained the exclusive right of steam navigation on the rivers and along the coast of that State; therefore, after Stevens had launched his boat on the Hudson, he was unable to employ it there. In this predicament he ventured on the hazardous experiment of taking his steam-vessel by sea, and successfully accomplished his voyage from New York to Delaware. This was the first attempt to put to sea in a steam-boat.

Mr. Stevens introduced many important improvements. He increased the length of stroke of the engines; he applied upright guides for the piston-rod, to supply the place of the parallel motion; and he divided the paddle-wheel by boards, by which means a more uniform motion was obtained. By these improvements he succeeded in raising the speed of steam-vessels to thirteen miles an hour.

Whilst Steam Navigation was making such progress in America, it was not neglected in this country. Mr. Henry Bell, of Glasgow, a man of great ingenuity, had for some time directed his attention to the subject, and had given some useful hints to Fulton. Seeing, as he afterwards said, no reason why others should profit by his plans without his participation in the fame and the profits, he determined to build a steam-boat himself, which was completed and launched in 1811. Bell called his boat the "Comet," in commemoration of the remarkable eccentric luminary which was at that time frightening Europe from its propriety. The boat was 25 tons burthen, with an engine of about 3-horse power. It plied on the Frith of Forth for a distance of 27 miles, which in ordinary weather it accomplished in 3½ hours. The "Comet" is generally supposed to have been the first steam-boat that plied regularly in Europe; and its construction was so perfect, that no boat built for many years afterwards surpassed it, taking into consideration its size and the small power of its engine. Bell, though he had done so much to advance Steam Navigation in this country, was allowed to suffer neglect and penury in his old age, till the town of Glasgow granted him a small annuity for his services.

A claim has been preferred on behalf of Messrs. Furnace and Ashton, of Hull, to priority in building the first steam-vessel that was worked in England. It is stated, that "about the year 1787, experiments were made on the river Hull, by Furnace and Ashton, on the propulsion of vessels by steam power. Furnace and Ashton built a boat, which plied on the river, between Hull and Beverley, for some time, and answered exceedingly well. In consequence of the good results of their experiments, they built a much larger vessel and engine, and sent the whole to London, to be put together and finished; after which it was subjected to the severest tests, and gave the greatest satisfaction. The vessel was bought by the Prince Regent (afterwards George IV.), who had it fitted and furnished as a pleasure yacht; but it was soon afterwards burnt, having, it is supposed, been wilfully set on fire by persons who were afraid that such an invention would be injurious to their calling. The Prince was so much pleased with the invention and ingenuity of Furnace and Ashton, that he granted them a pension for their lives of £70 a year each."* This steamer was on the paddle-wheel principle, propelled by a steam engine, to which was attached a copper boiler.

From this time forward the progress of Steam Navigation was very rapid. Steam-ships were built longer and larger, and with more powerful engines; and the most skilful builders rivalled each other in the construction and adaptation of their vessels and engines, so as to attain the highest possible speed. The locality in which Steam Navigation may be said to have had its birth continued for a long time to be pre-eminent, and steam-boats built on the Clyde still rank very high, if not the highest, in the scale of excellence.

The ordinary land steam engine required considerable alterations to adapt it to marine purposes; nor was it till great experience had been gained in propelling vessels by steam power, that the more essentially requisite modifications were adopted. It was found important, in the first place, to reduce the space

* British Association Report for 1853.

occupied by the machinery as much as possible. The boilers were consequently made of less dimensions, but more extensive in their heating surface. It was also found desirable to employ two engines instead of one, the pistons being made to rise and descend alternately. By this means the motion was rendered more equable, and by placing the cranks of the common shaft at right angles, the "dead points" were passed more readily, and the want of a fly wheel was thus compensated.

The steam-boats employed in this country were, almost from the first, and continue with few exceptions to be, on the low-pressure condensing principle; the whole of the machinery being placed below the deck. This renders it necessary to diminish the height of the engines as much as possible; and in all marine steam engines, till within the last twenty years, instead of having a working beam over the cylinders, a cross-head was placed at the top of the piston-rod, the action of which was conveyed by parallel motions to cross beams on each side, which were situated at the bottom part of each engine. The motion, compared with that of an ordinary land engine, was thus inverted. The proportions of the cylinders were also different; the length of stroke being shorter, to diminish the height, and the diameter consequently greater. The valves, and the gearing connected with them, the air pump, the condenser, and other subsidiary parts, do not differ essentially from those of land engines; but the governor is omitted, as it is found impracticable to work a marine engine with great regularity.

Latterly, many engineers have introduced, with much success, arrangements for communicating the action directly from the piston-rod to the crank, without the intervention of the beam and parallel motions. This is generally done by causing the piston-rod to work between guides, and a jointed arm connects it with the crank. One method of producing the same effect is to make the cylinders oscillate on pivots, as contrived by Mr. Murdoch, in the first model steam carriage, made in 1784. This principle has been successfully carried into operation by Messrs. Penn, of Greenwich. The oscillating cylinders accommodate themselves to the varying directions of the cranks, and the strain occasioned by guide rods is diminished; but when very large cylinders are required, the friction and the pressure on the pivots must tend to counterbalance the advantage otherwise obtained.

In the ordinary paddle-wheel steam-boats, the floats of the paddle-wheels are fixed at equal distances round the rim, radiating from the centre; therefore they enter and come out of the water obliquely. There is, consequently, a considerable loss of power attending the use of such paddle-wheels, as only one float at a time can be acting vertically on the water, and exerting the propelling force in a direct line. Several attempts have been made to remedy this defect, and to produce what is called "feathering" floats, every one of which will act against the water at right angles. The mechanism required for making this adjustment is, however, liable to get out of order, and the introduction of vertically acting floats has consequently been very limited.

The large projecting paddle-boxes are objectionable in sea-going ships, as they present so large a surface to the action of the wind, and either impede the course of the ship, or make it unweatherly. This inconvenience was experienced in the early progress of Steam Navigation, and many attempts were made to overcome it, by substituting a different kind of propeller. Recourse was had to the inventions of the ancients, from whom the paddle-wheel was taken, to find some other means of propulsion. A method of propulsion, similar in principle to the action of sculls at the back of a boat, had been contrived long before the inconvenience of paddle-wheels in Steam Navigation was experienced. In 1784, Mr. Bramah obtained a patent for a propeller similar in its forms to the vanes of a windmill, which by acting obliquely on the water as it revolved, pushed the boat forward. Ten years afterwards, an "aquatic propeller" was patented by Mr. William Lyttleton, a merchant in London. It consisted of a single convolution of a three-threaded screw, and may be considered to be the first screw propeller invented. Numerous other ingenious persons, among whom were Tredgold, Trevethick, Maceroni, and Millington, afterwards invented propellers on the screw principle; but none of them were sufficiently satisfactory in their results to come into practical use.

In 1836, Mr. Smith and Mr. Ericsson obtained a patent for a screw propeller, which nearly resembled Mr. Lyttleton's original contrivance; and by perseverance in struggling against the many obstacles with which he had to contend, Mr. Smith succeeded, though all previous efforts had failed. His partner, however, became disheartened by the obstacles thrown in their way, and left this country for America before the success of the screw was established.

The first ship fitted with the screw propeller was called the "Archimedes." It was a vessel of 237 tons burthen, with a draught of water of 9 feet 4 inches. The screw projected at the stern, and being turned rapidly round by the steam engine, the oblique action of the thread of the screw against the water impelled the vessel forward.

The "Archimedes" was originally fitted with a single-threaded screw, the threads of which were 8 feet apart, and there were two convolutions of the screw round the shaft. One convolution of the screw having been accidentally broken off, the ship was found to go faster in consequence; and, following the course of investigation suggested by the accident, Mr. Smith at last adopted a double-threaded screw, with only half a convolution. The average performance of the engines was 26 strokes per minute, and the number of revolutions of the screw in the same time was 138½. The "pitch" of the screw was 8 feet; that is, the space across one entire convolution of the thread would have measured 8 feet; consequently, had it been acting against a solid body, as a cork-screw when entering a cork, one revolution of the shaft would have advanced the vessel 8 feet, and the speed would have been 12½ miles an hour; but the utmost speed the "Archimedes" obtained was 9¼ nautical miles. The difference was owing to the screw "slipping" in the water, because the fluid yielded to the oblique action of the blades.

The results of the working of that experimental ship were so satisfactory, that other ships were soon built, with modifications of the form of the propeller. It was found disadvantageous to have an entire convolution of the thread of the screw; for one part of it worked in the wake of the other, and resistance was produced by the backwater. After numerous experiments, in which the dimensions of the screw were successively diminished, the propeller was at length

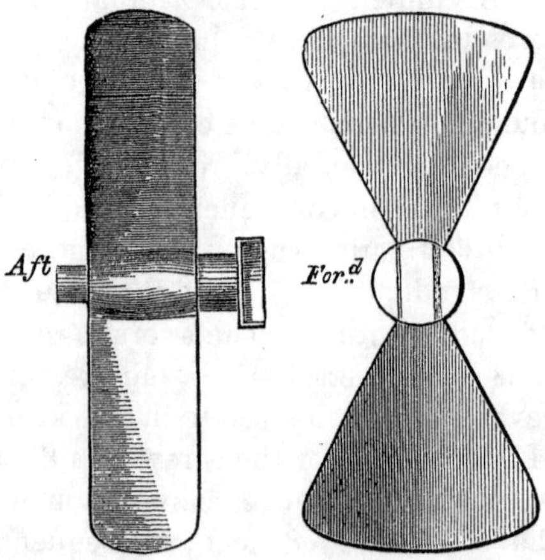

reduced to two oblique blades. Experiments on a large scale were conducted by Captain Carpenter, to determine the size and angle of inclination best adapted for the purpose of propulsion; and nearly all the ships now built for the Royal Navy are fitted with propellers on his principle. The annexed diagram represents on a scale of one-eighth of an inch to a foot, the form of the propeller of the "Agamemnon," of 606-horse power, which was recently engaged in successfully laying down the Atlantic Telegraph cable. The diameter of the screw is 18 feet, and the pitch 20 feet.

The screw propeller possesses great advantages in ships of war, as it is not exposed to damage by shot, and it leaves the entire deck clear for mounting guns. It has also the further advantage of not interfering with the working of sails, and is, therefore, admirably adapted for sea-going ships that economize fuel by alternately steaming and sailing, as the wind is adverse or favourable. The commotion in the water made by paddle-wheels, which is an objection to their use in narrow rivers, is avoided by screw propellers, which being immersed under the water, make little agitation on the surface, and the ships move along without any apparent impelling power.

The speed of ships with the best constructed screw propellers is fully equal to that of paddle-wheel vessels; and when two vessels of the same size, and with engines of equal power, one fitted with paddles, and the other with the screw, are fastened stem and stern together, in a trial of strength, the screw propeller has been found to have the advantage, and to pull its antagonist along at the rate of one or two miles an hour.

The difficulty at first experienced in the application

of the screw propeller was to communicate a sufficiently rapid motion to the shaft to which it is fixed; but, by the employment of direct-acting engines, this difficulty has been for the most part overcome. The power is generally first applied to drive a large cog-wheel, the teeth of which take into the teeth of a smaller cog-wheel fixed to the propeller shaft, and in this manner the velocity is sufficiently increased.

In 1852 the proportion of screw to paddle-wheel vessels building in the Clyde was as 43 to 30. The advantages of the propeller are becoming every year more appreciated, and it is rapidly superseding the paddle-wheel.

In the steam-boats of the United States the engines are constructed on the high-pressure principle; and by working with steam of the pressure of 100 pounds on the square inch, and with larger paddle-wheels, their boats attain a speed exceeding sixteen miles an hour. But numerous explosions of boilers on the North American rivers have operated as a caution against the introduction of high-pressure engines in steam-boats in this country. The dread of high-pressure steam was early impressed by the destructive explosion of the boiler of a steam-vessel at Norwich in 1817, which led to a long parliamentary inquiry into the subject; and the subsequent loss of life by the explosion of the "Cricket" on the Thames, has tended to strengthen the apprehension of high-pressure steam engines. For river use, however, when fresh water is always at command for generating the steam, there appears to be no more cause for fear of high-pressure engines in boats than on railways, provided the boilers are constructed with sufficient care. The experiments made by Mr. Fairbairn on the strength of boilers, the results of which were communicated at the meeting of the British Association in 1853, prove, that by increasing the number and strength of the "stays," or internal supports, of the boilers, they may be made, if sufficiently strong, to resist any possible pressure; and that the square shape, which was supposed to be the weakest, offers, on the contrary, peculiar facilities for giving increased strength. In one of these experiments made to determine the ultimate strength of the flat surfaces of boilers, when divided into squares of sixteen inches area, the boiler did not give way until it had sustained the enormous pressure of 1,625 pounds on the square inch.

It might be desirable, in the construction of steam boilers, to adopt the same principle that is introduced in the building of gunpowder mills, one-half of which is built in strong masonry, whilst the other is made of wood. By this means, when an explosion does occur, much less damage is done, for the lighter part only is blown away, which does little injury. In the same manner, steam engine boilers might be constructed with a small portion comparatively weaker, so that if it gave way there would not be much damage done. Safety-valves are intended to act in that manner; and if they were properly constructed, they would sufficiently answer the purpose, and guard against the possibility of danger; but the numerous accidents that occur with boilers provided with imperfect safety-valves, show that there is a necessity for some more effectual protection. Engineers are not sufficiently alive to the importance of improvements in this respect. They supply an engine with safety-valves, which would answer the purpose if kept in proper condition; but they do not make effectual provision against careless management and reckless misconduct. Some years since, a gentleman in America sent to the author a description, with drawings, of a safety-valve that combined the principles of the safety-plug without its inconvenience; it being so contrived that when the boiler became too hot, it melted some fusible metal which previously held down the valve, and then a weight pulled it open to allow an ample escape of steam; but when the heat was lowered, the valve again closed. This was shown to an eminent engineer for his opinion. He pronounced it to be very ingenious, and that it would, no doubt, answer the purpose; but he said, "*An improved safety-valve is not wanted*, those in use being quite sufficient for the purpose."

In steam-ships, where salt water is used for generating the steam, the incrustation on the sides of the boilers becomes a serious annoyance. It obstructs the communication of heat from the furnace to the water, and the metal is thus liable to become red-hot. Numerous plans have been adopted for the purpose of preventing the accumulation of salt on the sides of the boiler, the most common of which is to allow the water, when saturated with saline matter, to escape, and then to fill the boiler afresh. Among other contrivances for effecting the same purpose, without the waste of heating power which the change of water occasions, is Mr. Hall's plan of condensing the steam in dry condensers, cooled externally, so that the distilled water may be used again and again. This plan though theoretically good, is not much adopted; for the condensation of steam cannot be so well accomplished by that means as when a jet of cold water is thrown directly into the condenser. The principle of the dry condenser has, however, been lately made available in a new kind of engine, wherein the combined action of steam and of spirit vapour is applied as the propelling power.

Steam-boats had been for many years in extensive use on the rivers and seas of Europe and America before it was thought practicable to make voyages in them across the Atlantic. At the meeting of the British Association at Liverpool in 1837, that subject was brought forward for consideration, and it was then attempted to be shown, by calculations of the quantities of coal requisite for such a voyage, that steam communication with America would not be profitable, if it could be accomplished, as the coal would occupy so much of the tonnage as to leave scarcely any space for passengers and goods. Within a few months afterwards those calculations were set at nought by the "Sirius" and the "Great Western," which successfully crossed the Atlantic with passengers and cargo, the former in nineteen days from Cork, and the latter in sixteen. At the present time, steam-packets are constantly crossing from New York to Liverpool in eleven days.

Steam-ships now find their way to India and even to Australia, though the necessity of taking in coals at depôts supplied from England not only prolongs the time, but adds so materially to the cost, as to render steam communication with those distant places scarcely practicable with profit, since no freight can pay for the expense of coaling under such circumstances. To overcome that difficulty, it was proposed to build ships large enough to carry a supply of coals sufficient for the voyage there and back. One of those ships has been built for the Eastern Steam Navigation Company by Mr. J. Scott Russell, from the plans of Mr. Brunel, which is 675 feet long, 83 feet broad, and 60 feet deep. It is adapted to carry 6,000 tons burthen, in addition to the engines and requisite quantity of fuel, and to accommodate 2,000 passengers. This monster ship has been built on what is called the "wave principle" of ship-building, with long concave bows. It is to be propelled by the combined powers of the paddle-wheel and the screw. The engines for the former consist of 4 oscillating cylinders, 16 feet long and 74 inches in diameter, and the screw is to be worked by 4 separate engines, with cylinders of 84 inches in diameter. The speed which the "Great Eastern" is estimated to attain is 24 miles an hour, and it is calculated that the voyage to Australia will be accomplished in 30 days. There seems, at present, but small prospect of those calculations being realized, for the great cost incurred in launching the vessel and other expenses have exhausted the funds of the company by whom the ship was constructed.

Another company has, however, been formed for the purpose of completing, if possible, this great experiment in Steam Navigation; and the opinion so strongly expressed by Mr. Fairbairn at the recent meeting of the British Association at Leeds, of the strength of the monster ship, will give additional stimulus to their exertions. The ship is built on the same principle of construction as the Britannia Bridge over the Menai Straits, and it was stated by Mr. Fairbairn that it might be supported out of water, either in the centre or at each end, without injury.

STEAM CARRIAGES AND RAILWAYS.

No invention of the present century has produced so great a social change as Steam Locomotion on railways. Not only have places that were formerly more than a day's journey from each other been made accessible in a few hours, but the cost of travelling has been so much reduced, that the expense has in a great degree ceased to operate as a bar to communication by railway for business or pleasure.

Though the coaching system in this country had attained the highest degree of perfection, a journey from London to Liverpool, previously to the formation of railways, was considered a serious undertaking. The "fast coach," which left London at one o'clock in the day, did not profess to arrive in Liverpool till six o'clock the following evening, and sometimes it did not reach there till ten o'clock at night; and the fare inside was four guineas, besides fees to coachmen and guards. The same distance is now performed in six hours, at one-third the expense, and at one-fourth the fatigue and inconvenience.

Railway Locomotion, however, forms no exception

to the rule, that most modern inventions have their prototypes in the contrivances of ages past. They were used upwards of two hundred years before locomotive engines were known, or before the steam engine itself was invented. The manifest advantage of an even track for the wheels long ago suggested the idea of laying down wood and other hard, smooth surfaces for carriages to run upon. They were first applied to facilitate the traffic of the heavily laden waggons from the coal pits; the "tramways," as they were called, being formed of timber about six inches square and six feet long, fixed to transverse timbers or "sleepers," which were laid on the road. These original railways were made sufficiently wide for the wheels of the waggons to run upon without slipping off; the plan of having edgings to the rails, or flanges to the wheels, not having been adopted till a later period. To protect the wood from wearing away, broad plates of iron were afterwards fixed on the tramways.

Cast iron plate rails were first used in 1767. The flat plates on which the wheels ran were made about three inches wide, with edges two inches high, cast on the near side, to keep the wheels of the "trams" on the tracks. These iron plates were usually cast in lengths of six feet, and they were secured to transverse wooden sleepers by spikes and oaken pegs. The tramways were laid down on the surface of the country without much regard to hills and valleys, the horses that drew the trains being whipped to extra exertion when they came to a hill, and in descending some of the steep inclines, the animals were removed, and the loaded waggons were allowed to descend the hills by their own gravity, the velocity being checked by a break put on by a man who accompanied them.

The chief use of the tramways was to facilitate the conveyance of coals from the pits to the boats; and as the level of the pit's mouth was higher than that of the water, it was an object, in laying down a tramway, to make a continuous descent, if possible, for the loaded trains to run down, the dragging back of the empty ones being comparatively easy. Thus, though "engineering difficulties" were not much considered in the construction of those early railways, engineering contrivances were adopted to diminish the draught, by making the gradients incline in one direction.

Soon after the invention of the Steam Engine had been practically applied to mining purposes, its power was directed to draw the coal waggons on railways. This was done about the year 1808; and, in the first instance, the application of steam power was limited to drawing the loaded waggons up steep inclines. A stationary engine was erected at the top of the incline, and the waggons were drawn up by a rope wound round a large drum. This mode of traction was afterwards extended, in many instances, along the whole railway, so as to supersede the use of horse power. The employment of stationary engines in this manner was continued, even after the invention of locomotive steam engines, to draw the trains up inclines that were too steep for the power of the small locomotives at first used to surmount; nor has this plan been yet altogether abandoned.

The application of steam to the direct propulsion of carriages was a comparatively slow process. It was, indeed, contemplated by Watt, as a substitute for horse power on common roads, though he does not seem to have contrived any means by which it might be done. The first known application of the kind was made by Mr. Murdoch, an engineer in the employment of Messrs. Boulton and Watt, who in 1784 constructed a working model of a steam carriage, still preserved, and which formed one of the most interesting objects in the Great Exhibition of 1851. The boiler of this model locomotive is made of a short length of brass tube, closed with flat ends. The furnace to generate the steam consists of a spirit lamp. The steam is conducted directly from the boiler to a single cylinder, which is mounted on a pivot near the centre, so that by the movement of the cylinder the piston-rod may adapt itself to the varying positions of the crank. The two hind wheels are fixed to the axle, and on the latter is the crank, attached to the piston-rod. A single wheel in front serves to guide the carriage, which is propelled by the rotation of the two hind wheels. The elastic force of the steam is directly applied as the moving power; and after it has done its work in the cylinder, it is allowed to escape into the air.

This first known application of steam as a locomotive power is more perfect in its general arrangements than many steam carriages that were subsequently brought into operation; and in the plan of balancing the cylinder on pivots, we perceive the origin of the oscillating engines, which have been recently introduced with much success in Steam Navigation. By that arrangement there is attained the most direct application of the piston-rod to the crank, with the least loss of power.

Mr. Murdoch's intention was to employ such carriages on common roads, but he did not proceed to put his plan into operation. Several other engineers, among whom was Symington—who, as we have before seen, took an active part in the invention of Steam Navigation—afterwards endeavoured to realize Mr. Murdoch's ideas on a working scale; but the first who succeeded in making a locomotive engine, that ran with any success, were Messrs. Trevethick and Vivian. In 1804 they constructed a locomotive engine, which was employed on a mineral railway at Merthyr Tydvil, in South Wales. The boiler of their engine resembled the one in Mr. Murdoch's model, in having circular flat ends; but, to increase the heating surface, a flue was introduced in the middle of the boiler, which passed through it and back again, in the shape of the letter U. The lower part of the tube formed the furnace, and the upper part returned through the boiler into the chimney. The steam was admitted into and escaped from the cylinder by the working of a four-way cock, the contrivance of the slide-valve being then unknown. On the axle of the crank a cog-wheel was fixed, and, by means of the usual gearing, it communicated motion to the hind wheels, which were fixed to the axle, so that when the wheels revolved the carriage was propelled.

It is a remarkable fact that this engine of Mr. Trevethick's presents the first practical application of high-pressure steam as a motive power. Watt had, indeed, suggested the application of the impulsive power of steam, and Mr. Murdoch's model locomotive was necessarily constructed on that principle; but until Mr. Trevethick's locomotive engine was in action, no application of high-pressure steam had been made on a working scale.

The projectors of locomotive engines were for many years possessed with the notion that it was necessary to have some contrivance to prevent the wheels from slipping on the road, as it was supposed that otherwise the wheels would be turned without moving the carriage. Numerous plans were devised for overcoming this imaginary difficulty; and though experience proved that even on railways the adhesion of the wheels was, in ordinary circumstances, sufficient, yet various schemes continued to be tried for the purpose of facilitating the ascent of hills. The imitation of the action of horses' hoofs was one of the means attempted, but such additional aids were eventually found to be of no avail, and were discontinued.

All the endeavours that were made, in the first instance, to apply steam power to locomotion, had in view the propulsion of carriages on common roads, the idea of constructing level railways through the country, for facilitating the general traffic, being looked upon as too visionary a project to be realized. The inventors of locomotive engines consequently directed their attention almost exclusively to the arrangement that would best apply steam power to overcome the varying obstacles and undulations of common roads.

It is very curious and interesting, in tracing the progress of an invention, to observe the different phases through which it has passed, before it has been brought into the state in which it is ultimately applied. It not unfrequently happens that the original purpose sinks into insignificance, and is almost lost sight of, as the invention becomes more fully developed. Other objects, that were not perceived, or were considered altogether impracticable, present themselves, and are then pursued; and the invention, when perfected, is very different from its original design. Thus the endeavours of the first inventors of Steam Navigation were confined to the construction of steam-tugs that would propel the boats along canals, or take a ship into harbour, the notion of fitting a steam engine into a ship to propel it across the sea not having been thought of. In the same manner, the invention of Steam locomotion on railways was either not contemplated in the first instance, or was considered very subordinate to the construction of carriages to be propelled by steam power on common roads.

Among the most successful of those engineers, who constructed steam carriages to run on roads, were Mr. Gurney, Mr. Birstall, Mr. Trevethick, Mr. Handcock, and Colonel Maceroni. Mr. Gurney was one of the first on the road. His steam carriage completed several journeys very successfully, and proved the practicability of employing steam power in locomotive engines many years before the first passenger railway was brought into operation. This, like all other new inventions, was, however, beset with difficulties, among which the most annoying was the determined obstruction the plan met with from the trustees of public roads, who levied heavy tolls on the carriages, and laid loose stones on the roads to stop them from running, as the driving wheels were found to be destructive to the roads. There was also considerable danger in running steam carriages on the same roads on which ordinary traffic was conducted, because the strange appearance of the engines, their noise, and the issuing steam, frightened the horses.

Notwithstanding these difficulties, the importance of applying steam as a locomotive power for passenger traffic became so apparent, that a Committee of the House of Commons was appointed in 1831, to consider whether the plan could be adopted with safety on common roads, and whether it should not be encouraged by passing an Act of Parliament for regulating the tolls chargeable on such carriages, and for preventing the obstructions to which they had been exposed. The evidence given before the Committee was greatly in favor of steam carriages, and tended to show that there was no insuperable difficulty to the general adoption of them. The Committee accordingly reported as follows:—

"Sufficient evidence has been adduced to convince your Committee—

"1st. That carriages can be propelled by steam on common roads at an average speed of ten miles an hour.

"2nd. That at that rate they have conveyed upwards fourteen passengers.

"3rd. That their weight, including engines, fuel, water, and attendants, may be under three tons.

"4th. That they can ascend and descend hills of considerable elevation, with facility and safety.

"5th. That they are perfectly safe for passengers.

"6th. That they are not (or need not be, if properly constructed) nuisances to the public.

"7th. That they will become a speedier and cheaper mode of conveyance than carriages drawn by horses.

"8th. That as they admit of greater breadth of tire than other carriages, and as the roads are not acted upon so injuriously as by the feet of horses in common draught, such carriages will cause less wear of roads than coaches drawn by horses.

"9th. That rates of toll have been imposed on steam carriages which would prohibit them being used on several lines of roads, were such charges permitted to reman unaltered."

In defiance of this favourable report, experience proved that there were defects in that system of locomotion greater than its advocates were disposed to admit, and that the mechanism was frequently broken or disarranged by the constant jarring caused by the roughness of the road. The alarm of the horses drawing other carriages was also calculated to produce fearful accidents.

So far, indeed, as regarded the power of locomotion, the steam carriages were successful. The author was witness of this success during a short excursion in Colonel Maceroni's carriage, which ascended hills and ran over rough roads with great ease, and at a speed of twelve miles an hour. The practical difficulties, however, were so great, that steam carriages have not been able to compete with horse power; for the original cost of the boiler and engine, the necessary repairs, and the expense of fuel, amounted to more than the cost and keep of horses. The plan was practically tried for several weeks, in 1831, by running a steam carriage for hire from Paddington to the Bank of England. The carriage, of which the annexed diagram is an outline, was one of those

constructed by Mr. Handcock. The engine was placed behind the carriage, which was capable of containing sixteen persons, besides the engineer and guide. The latter was seated in front, and guided the carriage by means of a handle, which turned the fore wheels. The carriage was under perfect control, and could be turned within the space of four yards. With this carriage, Mr. Handcock stated he accomplished one mile up hill at the rate of seventeen miles an hour. The carriage loaded very well at fares which would now be considered exorbitant, but the frequent necessity for repairs rendered the enterprise unsuccessful, and the steam carriage was taken off the road.

The successful establishment of railways, and the great advantages arising from them compared with the ordinary means of conveyance, still further reduced the chance of establishing Steam Locomotion on roads, and the plan is now in abeyance, at least, if it has not been abandoned. It is very possible, however, that in the progress of invention, modifications may be made in the steam engine, to adapt it more successfully to the purpose; or more suitable motive powers may be discovered, that may bring mechanical locomotion on roads again into favour.

The successful application of Steam Locomotion on railways cannot be dated more than thirty years ago; yet in that short period its progress has been so rapid, that but few traces of the old mode of travelling by stage coaches are now to be seen.

Some locomotive steam carriages had, indeed, been introduced on the Stockton and Darlington coal railway, by Mr. George Stephenson, in 1825, but their results were not so satisfactory as to induce the extension of the plan to the other railways that were then laid down in the coal districts of England. The cylinders of those engines were vertical, and each of the four wheels acted propulsively on the rails by means of an endless chain running along cog-wheels fixed on the axles. The utmost speed that could be obtained by this means was eight miles an hour; and so little were these engines calculated to solve the problem of the practicability of steam locomotive engines, that when the first passenger railway was projected, from Liverpool to Manchester, it was proposed to propel the carriages by the traction of ropes, put in motion by stationary steam engines. The directors, before finally determining on the system of locomotion to be adopted, offered a premium of £500 for the best locomotive engine to run on that line. The stipulations proposed, and the conditions which the required engines were to fulfil, may be regarded as a curious exposition of the limited views then taken of the capabilities of Steam Locomotion on railways. The engine "was to consume its own smoke; to be capable of drawing three times its own weight at 10 miles an hour, with a pressure on the boiler not exceeding 50 pounds on the square inch; the whole to be proved to bear three times its working pressure—a pressure guage to be provided; to have two safety-valves, one locked up; the engine and boiler to be supported on springs, and rested on six wheels, if the weight should exceed 4½ tons; height to the top of the chimney not to exceed 15 feet; weight, including water in boiler, not to exceed 6 tons, or less, if possible; the cost of the engine not to exceed £550."

An engine, called the "Rocket," constructed by Messrs. Booth and Stephenson, was the successful competitor for the prize. It so far exceeded the required conditions as to speed, that, when unattached to any carriages, it ran at the rate of 30 miles an hour. The principal cause of its successful action was the introduction of a boiler perforated lengthwise by many tubes, through which the heated air of the furnace passed to the chimney, and by this means a much larger evaporating surface was obtained than in the boilers previously employed, with a single flue passing through the centre. The tubes were of copper, three inches in diameter, one end of each communicating with the chimney, and the other with the furnace. There were twenty-five of these tubes passing through the boiler, and fixed water-tight at each end.

The boiler was 3 feet 4 inches in diameter, and 6 feet long; and it exposed a heating surface of 117 square feet. There were two cylinders, placed in a diagonal position, with a stroke of 16½ inches, and each worked a wheel 4 feet 8½ inches diameter, the piston-rod being attached externally to spokes of the driving wheels. The draught of the chimney, aided by the escaping steam from the cylinders, which was admitted into it, served to keep the fuel in active combustion. The "Rocket" weighed 4¼ tons; the tender, with water and coke, 3 tons 4 cwt.; and two loaded carriages attached, 9½ tons; so that the engine and train together weighed about 19 tons. The boiler evaporated 114 gallons of water in the hour, and consumed, in the same time, 217 pounds of coke. The average velocity of the train was 14½ miles per hour.

The accompanying woodcuts represent an elevation of the "Rocket," and a section of its boiler. In these figures, *a* is the fire-box or furnace, surrounded on all sides with water, with the exception of the side perforated for the reception of the tubes; *b* is the boiler; *d*, one of the steam cylinders; *e*, the chimney; *h* and *i*, safety-valves; *f*, one of the connecting rods for communicating motion to the driving wheels.

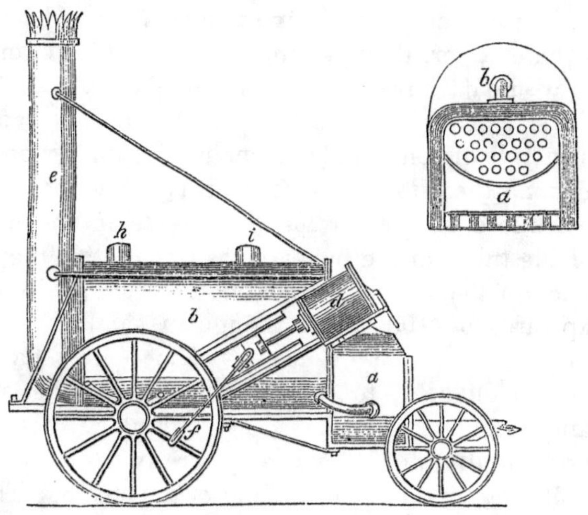

Three other engines competed with the "Rocket," two of which had attained great speed on previous trials. These were the "Novelty," constructed by Messrs. Braithwaite and Ericsson, which weighed only 2¾ tons; and the "Sans Pareil," manufactured by Mr. Arkworth, which weighed 4½ tons. On the day of trial, the 6th of October, 1829, these two locomotive engines were disabled by the bursting of some of their pipes, and thus the field was left clear to the "Rocket," for the fourth engine had no chance of winning the prize.

The "Rocket," indeed, more than fulfilled all the conditions required by the directors of the railway, who thereupon decided on employing locomotive engines for the traffic on the line.

The "Rocket" has formed the model on which all subsequent locomotive engines have been constructed; for, though numerous alterations and improvements have been made in details, and though the size of the engines has been greatly enlarged, the principle of construction remains essentially the same. Among the improvements that have been introduced by different inventors, is an increase in the number of the tubes in the boiler, so as to facilitate the generation of steam, some of the engines now made having upwards of 100 tubes, though of smaller diameter than those of the "Rocket." The boilers have also been elongated, to enlarge the evaporating surface and economize fuel. The cylinders are placed horizontally, and they are generally fixed inside the boiler, to prevent the cooling of the steam. The piston-rods are attached to cranks on the axle, placed at right angles to each other; and the engines are generally mounted on six wheels, four of which are driving wheels, made of larger size than the two others, and they are coupled together by connecting arms. The large and powerful engines on the Great Western Railway have, however, only two driving wheels, which are 8 feet in diameter. These engines weigh as much as 31 tons, which is seven times more than the weight of the "Rocket." They are capable of taking a passenger train of 120 tons at an average speed of 60 miles an hour on easy gradients; and the effective power, as measured by a dynamometer, is stated to be equal to 743 horses.

The accompanying engraving of one of the recently constructed engines on the Great Western Railway presents a remarkable difference in point of size and general arrangement to the original prototype, from which, however, it does not materially differ in the principle of its construction.

The complete success of the "Rocket" having settled the question of the mode of traction, the Directors of the Liverpool and Manchester Railway made increased efforts to complete the line, and to open it for general traffic. In September, 1830, all was ready for the opening, which it was determined should take place with a ceremony indicative of the importance of the great event. The principal members of the Government consented to take part in the inauguration of the railway, and the utmost interest was excited throughout the country for the success of an undertaking that promised to be the commencement of a new era in travelling. The 15th of September was the day appointed, and there were eight locomotive engines provided to propel the same number of trains of carriages, which were to form the procession. All along the line there were crowds of persons collected to witness the ceremony. The trains started from the Liverpool end of the railway ; and, as they passed along, they were greeted by the cheers of the astonished and delighted spectators. On arriving at Parkside, seventeen miles from Liverpool, the engines stopped to take in fresh supplies of fuel and water. The passengers alighted and walked upon the line, congratulating one another on the delightful treat they were enjoying, and on the success of the great experiment. All hearts were bounding with joyous excitement, when a disastrous event occurred, which threw a deep gloom over the scene. The Duke of Wellington, Sir Robert Peel, and Mr. Huskisson were among those who were walking on the railway, when one of the engines was recklessly put in action, and propelled along the line. There was a general rush to the carriages, and Mr. Huskisson, in trying to enter his carriage, slipped backwards and fell upon the rails. The wheels of the engine passed over his leg and thigh, and he was so severely injured, that he expired in a few hours.

Notwithstanding this lamentable occurrence, the journey was continued to Manchester, and the carriages returned to Liverpool the same evening. On the following morning the regular trains commenced running, and they were crowded with passengers, nothing daunted by the fatal calamity on the opening day.

The immense advantages of this mode of travelling were at once apparent, and lines of railway in different parts of the country were quickly projected. The railway from London to Birmingham was the first one commenced after the completion of the Liverpool and Manchester line, and a connecting link with Manchester and Liverpool was also begun by a separate company. The Birmingham Railway was opened throughout on the 17th September, 1838.

Railway enterprise was not checked by the great cost of the undertakings, nor by the miscalculations of the engineers, who, in the first instance, frequently greatly under-estimated the expenditure requisite for the cuttings, embankments and tunnels, which were thought necessary to attain as perfect a level as possible. The original estimate for the Liverpool and Manchester Railway was £300,000, but the amount expended on the works at the time of opening was nearly £800,000. The original estimate of the London and Birmingham Railway, including the purchase of land, and the locomotives and carriages, was £2,500,000, whilst the actual cost amounted to £5,600,000, the cost of the works and stations being about £38,000 per mile. The Grand Junction Railway, from Birmingham to Liverpool, was more economically constructed, because the difficulties to be surmounted were not so great, and less attention was paid to maintain a level line. It was estimated to cost, including all charges, £13,300 per mile, though the actual cost was £23,200.

The plan adopted for laying down and fixing the rails on all the railways in England, with the exception of the Great Western, is nearly similar to that on which the original coal-pit railways were constructed. Pieces of timber, called "sleepers," are laid at short distances across the road, and on to these sleepers are fixed cast iron "chairs," into which the rails are fastened by wedges, the sleepers being afterwards covered with gravel or other similar material, called "ballast," to make the timbers lie solidly, and to keep the road dry.

The railway system of Great Britain was commenced without sufficient attention to the determination of the best width apart of the rails. In forming the Liverpool and Manchester Railway, the guage of the railways in the collieries was adopted, and the width between the rails was made 4 feet 8½ inches. The same width of rails was adopted on the London and Birmingham and Grand Junction Railways; and as uniformity of guage was essential to enable the engines and carriages on one line to travel on another, the other railways connected with the grand trunk line were made of the same width of guage. Mr. Brunel, the engineer of the Great Western Railway, departed from that uniformity, and laid down the rails 7 feet apart. The increased width of guage possesses many advantages, of which greater steadiness of motion and greater attainable speed, without risk, are the most important; but, at the same time, the additional space incurs a greater expense in laying out the line. As branches from the Great Western Railway spread into the districts where the narrow guage railways had been laid down, much inconvenience has arisen from the break of guage, as it occasions the necessity for a change of carriages. On some railways, to avoid this inconvenience, narrow and broad guage rails have been laid down on the same line.

If the railway system of Great Britain were to be recommenced, after the experience that has now been acquired, the medium guage would most probably be adopted; and in commencing to lay down railways in Ireland, the Irish Railway Commissioners recommended 6 feet 2 inches as the most desirable width, and that standard has been advantageously adopted in the sister country.

Travelling experience tells greatly in favour of the broad gauge. There is no railway out of London whereon the carriages run so smoothly, and on which the passengers are so conveniently accommodated, as on the Great Western. The speed attained on that railway also surpasses that on any other. The express train runs from London to Bristol, a distance of 120 miles, in less than three hours. The author accompanied an experimental train, when one of the large engines was first put upon the line, and during some portion of the journey a rate of 70 miles an hour was accomplished without any inconvenient oscillation.

It must be observed, with regard to the action of locomotive engines, that as the piston-rods are attached directly to cranks on the axle, each piston makes a double stroke for every revolution of the driving wheels; consequently, when the engine is running at great speed, the movement of the piston is so rapid, that there is neither time for the free emission of the waste steam, nor for the full action of the high-pressure steam admitted. There is, therefore, a great waste of power occasioned by the admitted steam having to act against the steam that is escaping; and an engine, calculated to have the power of 700 horses, will not exert a tractive force nearly equal to that amount. With a driving wheel 6 feet in diameter, a locomotive engine will be propelled 18 feet by each double stroke of the piston, if there be no slipping on the rails; consequently, in the space of a mile, the piston must make 300 double strokes. When running, therefore, at the speed of 30 miles an hour, the piston makes 150 double strokes per minute.

The success of the great experimental railway from Manchester to Liverpool not only stimulated similar works in this country, undertaken by private enterprise; but the Continental Governments quickly perceived the importance of that means of communication, and commenced the formation of railways at the national cost, and placed them under governmental control. Belgium was peculiarly adapted, by the general level state of the country, for the formation of railways; and long before any connected system was completed in this country, the *chemins de fer* formed a complete net-work in that kingdom, and the system of conducting the traffic was brought to a much higher state of perfection than was attained in this country. The rate of travelling, however, was slower.

It is a question that has been often mooted, whether it is better to allow the system of communication throughout the country to be conducted by independent companies of enterprising individuals, or to place it entirely under the control of the Government. The want of system manifested in the formation of the railways in England has proved a serious inconvenience, and has occasioned wasteful expenditure, besides having led to a fearful destruction of life, owing to the want of careful attention to the means of safety, and to ill-judged parsimony in the management of the traffic. There can be no doubt that if the Government had undertaken the work zealously, and with the view of establishing a complete system of railway communication, many of the inconveniences now experienced might have been avoided, and the railways might have been laid down and worked at considerably less cost, and with a large addition to the national revenue. There is, however, so strong a disinclination in this country to the centralization of Government power, and to the extension of Government influence, that the people generally had rather submit to considerable inconvenience and expense, than tolerate the system of railway management which has been adopted on the Continent. The necessity of interference, to protect the interests of the public, has nevertheless compelled the Government, though late, to adopt measures for controlling the management of the railway companies, and stringent regulations are now imposed with a view to prevent unnecessary danger to railway passengers.

The railway system of Great Britain, though established entirely by private enterprise, represents an amount of capital equal to one-third of the national debt, and nearly 100,000 individuals are directly employed in conducting the traffic on the various railways in this kingdom. An idea of the vastness of these undertakings, and the important interests involved in them, may be formed from the following facts, stated by Mr. Robert Stephenson, at the Institution of Civil Engineers:—

"The railways of Great Britain and Ireland, completed at the beginning of 1856, extended 8,054 miles, and more than enough of single rails were laid to make a belt round the globe. The cost of constructing these railways had been £286,000,000. The working stock comprised 5,000 locomotive engines and 150,000 carriages and trucks; and the coal consumed annually by the engines amounted to 2,000,000 tons, so that in every minute 4 tons of coal flashed into steam 20 tons of water. In 1854 there were 111 millions of passengers conveyed on railways, each passenger travelling an average of 12 miles. The receipts during 1854 amounted to £20,215,000; and there was no instance on record in which the receipts of a railway had not been of continuous growth, even where portions of the traffic had been abstracted by new lines. The wear and tear of the railways was, at the same time, enormous. For instance, 20,000 tons of iron rails required to be annually replaced, and 26 millions of wooden sleepers perished in the same time. To supply this number of sleepers, 300,000 trees were felled, the growth of which would require little less than 5,000 acres of forest land. The cost of running was about fifteen pence per mile, and an average train will carry 200 passengers. Without railways, the penny post could not have been established, because the old mail coaches would have been unable to carry the mass of letters and newspapers that are now transmitted. Every Friday night, when the weekly papers are published, eight or ten carts are required for Post Office bags on the North-Western Railway alone, and would hence require 14 or 15 mail coaches."

Adverting to other advantages derived from railway locomotion, Mr. Stephenson noticed the comparative safety of that mode of travelling. Railway accidents occurred to passengers in the first half of 1854 in the proportion of only one accident to every 7,194,343 travellers. As regards the saving of time, he estimated that on every journey, averaging 12 miles in length, an hour was saved to 111 millions of passengers per annum, which was equal to 38,000 years, reckoning eight working hours per day; and allowing each man an average of 3s. a day for his work, the saving of time might be valued at £2,000,-

000 a year. There were 90,000 persons employed directly, and 40,000 collaterally, on railways; and 130,000 men, with their families, represent 500,000 so that 1 in 50 of the entire population of the kingdom might be said to be dependent for their subsistence on railways.

Every year adds to the extent of the railway system, and to the increase of the traffic, so that considerable addition should be made to the amounts stated by Mr. Stephenson to represent the state of railway enterprise and railway traffic at the present day. The traffic returns for the week ending the 25th of September, 1858, amounted to £502,720; and the gross receipts of the railways in 1857 were £24,174,-610. The railways now open for traffic in England, Scotland, and Ireland extended to upwards of 9,000 miles, and the lines reported to be in the course of construction amount to one-ninth the length of those completed.

In estimating the importance and advantage of railway travelling, there must not be omitted its cheapness and comfort, compared with travelling by stage coach. There are some persons, indeed, who look back with regret to the old coaching days; and it must be admitted that railways have taken away nearly all the romance of travelling, and much of the exhilarating pleasure that was experienced when passing through a beautiful country on the top of a well-horsed coach in fine weather. The many incidents and adventures that gave variety to the journey were pleasant enough for a short distance; but two days and a night on the top of a coach, exposed to cold and rain, or cramped up inside, with no room to stir the body or the legs, was accompanied with an amount of suffering which those who have experienced it would willingly exchange for a seat, even in a third-class railway carriage. In a national and in a social point of view, also, railways have produced important improvements. They tend to equalize the value of land throughout the kingdom, by bringing distant sources of supply nearer the points of consumption; they have given extraordinary stimulus to manufacturing industry; and by connecting all parts of the country more closely together, railway communication has concentrated the energies of the people, and has thus added materially to their wealth, their comforts, and to social intercourse.

Nor must we, in noticing the grand invention of locomotion on railways, omit to mention some of the many subsidiary works which have been created during its progress towards perfection, and which have contributed to its success. Tunnels, of a size never before contemplated, have penetrated for miles through hard rocks, or through shifting clays and sands; embankments and viaducts have been raised and erected, on a scale of magnitude that surpasses any former similar works; bridges of various novel kinds, invented and constructed for the special occasions, carry the railways over straits of the sea, through gigantic tubes; across rivers, suspended from rods supported by ingeniously devised piers and girders; and over slanting roads, on iron beams or on brick arches built askew. As to the locomotive engines, though the principle of construction remains the same, the numerous patents that have been obtained attest that invention has been active in introducing various improvements in the details of construction, to facilitate their working, and to increase their power. The various plans that have been contrived for improving the structure of the wheels and axles, for the application of breaks, for deadening the effect of collisions, for making signals, for the forms of the rails, and for the modes of fastening them to the road, are far too many to be enumerated.

In addition to the innumerable contrivances that have been invented for the improvement of the working of ordinary railways, several distinct systems of railway locomotion have been introduced to public notice, some of which seemed very feasible, though they have nearly all gradually disappeared. Of these, the Atmospheric railway was the most promising, and for a time it bid fair to supersede the use of locomotive engines. The propulsion of the carriages, by the pressure of the atmosphere acting on an attached piston working in a vacuum tube, possessed many theoretical advantages, and if it could be applied economically, railway travelling would become more pleasant and more free from danger than it is. On several lines of railway the atmospheric plan was put into operation, but owing to the expense of working, it was gradually abandoned. The short line from Kingston to Dalky, in Ireland, up a steep incline, was favourable to the working of the atmospheric railway, and there it continued to linger for some time after it had been abandoned elsewhere.

It is to be regretted that the atmospheric railway should have failed in economical working, for it possessed greater advantages for general traffic than the ordinary locomotive railway trains; and it is probable that if the same amount of inventive power and industry, which have been bestowed in improving locomotive engines, had been directed to overcome the difficulties of atmospheric traction, it might have proved economically successful.

The facility of travelling by railway has excited a spirit of locomotion before undreamed of. Instead of the diminished demand for horses which was apprehended when railways displaced stage coaches, public conveyances have increased a hundredfold. We can now scarcely conceive the time when there was not an omnibus in the streets of London, yet, scarcely more than thirty years ago, they were unknown, and travelling by stage carriages from one part of the town to another was prohibited by law! On their first introduction, omnibuses were considered absurdities, and were ridiculed as "painted hearses." The present omnibus traffic in London alone amounts to nearly £20,000 per week.

THE AIR ENGINE.

Numerous attempts have been made to supersede steam as a motive power, with the view to avoid the loss of heat by its absorption in the steam in a latent state. Mercury vapour and spirit vapour have been tried, in the expectation that as they possess much less capacity for heat, an equal pressure might be obtained, with a diminished loss of heating power. Several gaseous agents have been applied to the same purpose, of which carbonic acid gas seemed to present the best prospect of success, because it becomes expanded with a comparatively small increase of temperature. None of these attempts to produce a motive power superior to steam have yet proved successful. They have all, after a short season of promise, dropped out of notice; and the only one that is still in the field, struggling for superiority, is the air engine.

The first known air engine was invented by Sir George Cayley, in 1803. In his engine the air was heated by passing directly through the hot coals of the furnace, which some engineers yet consider to be the best mode of expansion; but its operation did not answer expectations. Mr. D. Stirling, of Dundee, afterwards improved on Sir George Cayley's plan, and introduced a method of regaining the heat from the expanded air, after it had done its work in the cylinder, and of applying it to expand the air again. Engines on this construction have been for some years working in Scotland, and in 1850 Mr. Stirling took out a patent for an improvement in the arrangement, which is stated to have been very successful.

Though Sir George Cayley and Mr. Stirling were the first in the field as inventors of air engines, the name of Mr. Ericsson, an American, is more closely associated with the invention, as he has for many years been conducting experiments on a large scale, and has tried his "caloric engine" on land, and on a ship of large burthen, built for the purpose.

The principle and the working of Mr. Ericsson's caloric engine is nearly the same as Mr. Stirling's; but as it has been brought most prominently into notice, we shall direct attention more particularly to its construction and performances. Mr. Ericsson obtained a patent for his caloric engine in this country in 1833, and a subsequent patent for improvements on it was taken out in 1851. During those years, and to a late period, he was indefatigably working out the principle, and numerous highly favourable reports have from time to time been made of the results of the experiments; but the advantages to be derived from the air engine remain nevertheless very questionable.

The object attempted to be gained is to make the same heating power do its work again and again. Atmospheric air, after being expanded by passing over an extensive hot surface, exerts the force thus acquired to raise the piston of a large cylinder, and it is then attempted to abstract the heat as the air issues out, and to apply it to the expansion of a further quantity.

The practicability of this plan has undergone much discussion; its friends and foes being equally confident in their opinions. The former pronounce it to be one of the most valuable inventions of the age, being calculated to economize heat, and to give greatly additional impulse to navigation; whilst its opponents declare that the calculations are erroneous, the experiments fallacious, and that the expanded air consumes more heating power than steam.

In one of the favourable notices of Mr. Ericsson's engine in an American publication, it is thus described:—"Two caloric engines have been constructed in New York, one of 5-horse power, the other of 60. The latter has four cylinders; two of 6 feet diameter, placed side by side, surmounted by two of much smaller size. Within are pistons, so connected that those in the lower and upper cylinders move together. A fire is placed under the bottom of the large cylinders, called the working cylinders; those above are called the supply cylinders. As the piston in the supply cylinder moves down, valves at the top admit the air. As they rise, those valves close, and the air passes into a receiver and regenerator, where it is heated to about 450°, and entering the next working cylinder, it is further heated by a fire underneath to 485°. The air is thus expanded to double its volume; and supposing the supply cylinder to be half the size of the other, the air, when expanded, will completely fill the larger cylinder. As the area of the piston of the smaller cylinder will be only half that of the larger, and as the air will be of the same pressure in both, the total pressure on the piston of the large cylinder will be double that on the small one. This surplus furnishes the working power of the engine. After the air in the working cylinder has forced up the piston within it, a valve opens; and as the air passes out, the piston descends by gravity, and cold air rushes in, and fills the supply cylinder.

"The most striking feature is the regenerator. It is composed of wire net, placed together to a thickness of about 12 inches. The side of the regenerator, near the working cylinder, is heated to a high temperature. The air passes through it before entering the working cylinder, and becomes heated to 450°. The additional heat of 30° is communicated by the fire underneath to the large cylinder. The expanded air forces the cylinder upwards, valves open, and it passes from the cylinder, and again enters the regenerator. One side of the regenerator is kept cool by the air on its entering in the opposite direction at each stroke of the piston; consequently, as the air of the working cylinder passes out, the wires abstract its heat so effectually, that when it leaves the regenerator, it has been robbed of all except about 30°. In other words, as the air passes into the working cylinder, it gradually receives from the regenerator about 450° of heat; and as it passes out, this is returned to the wires, and it is thus used over and over again;

the only purpose of the fires beneath the cylinders being to supply the 30° of heat which are lost by radiation and expansion.

"The regenerator in the 60-horse engine measures 26 inches in height and width. Each disc of wire composing it contains 676 superficial square inches, and the net has 10 meshes to the inch. Each superficial inch, therefore, contains 100 meshes, and there are 67,600 in each disc; and as 200 discs are employed, the regenerator contains 13,520,000 meshes, with an equal number of small spaces between the discs as there are meshes; therefore, the air is distributed into 27,000,000 of minute cells. The wire in each disc is 1,140 feet long; and the total length of wire in the regenerator is 41½ miles, or equal to the surface of four steam boilers, each 40 feet long and 4 feet diameter."

The accounts received from America of the great success that had attended the working of Mr. Ericsson's air engine,on the ship "Ericsson," attracted much attention in this country, and formed the subject of two evenings' discussion in the Institution of Civil Engineers. The most prevalent opinion was, that it is impossible to regain the heating power without corresponding loss of mechanical force or the addition of heat, and that there must have been some fallacy in the reports of the work done and of the quantity of fuel consumed.

It is, indeed, evident that nothing approaching the amount of heat said to have been recovered could be regained by passing through the regenerator; for as the apparatus becomes heated by the first portions of air passing through it, the temperature of the quantity that afterwards passed must at least be equal to that of the heated wires, and the last portions of air would consequently scarcely part with any caloric to the regenerator, previously heated to nearly its own temperature. Experience has since proved that the notion of regaining the heat by the regenerator was fallacious, for in the last improvements in Mr. Ericsson's engine, it is stated that the regenerator has been abandoned, and the plan has been adopted of cooling the air as it issues from the large cylinder, by passing it through tubes surrounded by cold water, and then using the same air over again.

One great practical inconvenience in the use of the air engine was the necessity of having enormously large cylinders to attain the required power, with the low amount of pressure that can be procured by the expansion of the air. The consequent friction increased the loss of power, and the difficulty of lubricating the pistons added to the practical objections to the air engine. To overcome these objections, the air in Mr. Stirling's engine is compressed before it is heated, by which means an equal amount of pressure is obtained on a smaller piston.

The air engine would in many respects possess advantages over the steam engine, if it could be worked economically. The space occupied by the boilers would be saved, and the danger of explosions would be avoided; for hot air does not scald, and the quantity at any time expanded would be too small to do much injury.

A patent has since been obtained by Messrs. Napier and Rankine, for improvements in the air engine, which they anticipated would remove the objections that have been raised to the engines of Stirling and Ericsson. The heating surface has been greatly increased by employing tubes; and other defects in the former engines, to which their want of complete success is attributed, have been remedied, so that Mr. Rankine, in his description of the improvements at the meeting of the British Association at Liverpool, confidently anticipated to effect a great saving of heating power, combined with the other advantages of the air engine. He estimated the consumption of fuel by a theoretically perfect air engine on Mr. Stirling's principle at 0·37 lbs. per horse power per hour; whilst a theoretically perfect steam engine would consume 1·86 lbs. The actual average consumption of a steam engine is, however, 4 lbs. of fuel per horse power per hour, and the actual consumption of Stirling's engine is stated by Mr. Rankine to have been 2·20 lbs, and that of Ericsson's 2·80 lbs. It appears from this statement, therefore, that the air engines of Messrs. Stirling and Ericsson are superior in point of economy of fuel to steam engines; and if Mr. Rankine's anticipations of the superiority of his air engine be realized, it will effect still greater economy. In Messrs. Napier and Rankine's engine, the air is compressed before expansion, so that the size of the cylinders may be reduced to even smaller dimensions than the cylinders of steam engines of equal power.

PHOTOGRAPHY.

THE power we now possess of fixing the transient impression of the rays of light, and of retaining the beautiful images of the camera obscura, is perhaps the most astonishing of the present age of wonders. Effects similar to those of the electric telegraph, of steam navigation, of dissolving views, and of other wondrous realizations of inventive genius, had been anticipated in growing tales of Eastern romance centuries ago; but the most fanciful imagination had not conceived the possibility of making Nature her own artist, and of producing, in the twinkling of an eye, a permanent representation of all the objects comprehended within the range of vision.

Such an idea could scarcely have occurred until after the invention of the camera obscura; but when looking at the beautiful pictures focused on the screen of that instrument, it became an object of longing desire to fix them there.

To trace the history of Photography from its earliest beginnings, we must go back to the days of the alchemists, who were the discoverers of the influence of light in darkening the salts of silver, on which all photographic processes on paper depend. That property of light was noticed in 1566, and it induced the speculative philosophers of that day to conceive that luminous rays contained a sulphurous principle which transmitted the forms of matter. Homberg, more than a century afterwards, misled by this action of the sun's rays, supposed that they insinuated themselves into the particles of bodies, and increased their weight; and Sir Isaac Newton also entertained a similar opinion.

The influence of the solar rays in facilitating the crystallization of saltpetre and sal ammoniac, was shown by Petit in 1722; and in 1777, the distinguished chemist Scheele discovered that the violet rays of the spectrum possess greater power in producing those changes than any other. A solution of nitrate of silver, then called "the acid of silver," was known to be peculiarly susceptible to the action of those rays. The experiment by which it was illustrated consisted in pouring the solution on chalk, which became blackened by exposure to light. These discoveries were made by Scheele in his endeavours to find in light the source of "phlogiston" —that *ignis fatuus* of the chemists of the last century. We thus perceive, in the first steps towards the invention of Photography, one of the many instances of the discovery of truth in the search after error.

At the beginning of the present century, Mr. Wedgwood, the celebrated porcelain manufacturer, undertook a series of experiments to fix the images of the camera, assisted by Mr. (afterwards Sir Humphry)

Davy. They so far succeeded as to impress the images on the screen, but unfortunately they had not the power of preserving the paper from being blackened all over when exposed for a short time to the light. "Nothing," said Sir Humphry Davy, in his account of these experiments, "but a method of preventing the unshaded parts of the delineation from being coloured by exposure to light is wanting to render this process as useful as it is elegant."

It was in June, 1802, that Mr. T. Wedgwood published "an account of a method of copying paintings on glass, and of making profiles by the agency of light; with observations by H. Davy." Mr. Wedgwood made use of white paper or white leather, moistened with a solution of nitrate of silver. The following description of the process, contributed to the "Journals of the Royal Institution" by Davy, will be read with interest, as showing how closely these experiments approximated to the photogenic process, invented by Mr. Talbot thirty-six years afterwards :—

"White paper or white leather moistened with a solution of nitrate of silver undergoes no change in a dark place; but on being exposed to daylight, it speedily changes colour, and after passing through different shades of grey and brown, becomes at length nearly black; the alterations of colour take place more speedily in proportion as the light is more intense. In the direct rays of the sun, two or three minutes are sufficient to produce the full effect. In the shade, several hours are required; and light transmitted through different coloured glasses acts on it with different degrees of intensity. Thus it is found that red rays, or the common sunbeams passed through red glass, have very little action on it. Yellow or green are more efficacious; but blue and violet light produce the most decided and powerful effects.

"When the shadow of any figure is thrown on the prepared surfaced, the part concealed by it remains white, and the other parts speedily become dark. For copying paintings on glass, the solution should be applied on leather, and in this case it is more readily acted on than when paper is used. When the colour has been once fixed on leather or paper, it cannot be removed by the application of water, or water and soap, and it is in a high degree permanent. The copy of a painting or a profile, immediately after being taken, must be kept in a dark place. It may, indeed, be examined in the shade, but in this case the exposure should only be for a few minutes; by the light of candles or lamps, it is not sensibly affected. No attempts that have been made to prevent the uncoloured parts of the copy or profile from being acted upon by light, have as yet been successful. They have been covered with a coating of fine varnish, but this has not destroyed their susceptibility of becoming coloured; and even after repeated washings, sufficient of the active part of the saline matter will still adhere to the white parts of the leather or paper, to cause them to become dark when exposed to the rays of the sun.

"The woody fibres of leaves, and the wings of insects, may be pretty accurately copied; and in this case it is only necessary to cause the direct solar light to pass through them, and to receive the shadows on prepared leather. Images formed by means of the camera obscura have been found too faint to produce, in any moderate time, an effect on nitrate of silver. To copy those images was the first object of Mr. Wedgwood in his researches on this subject, and for this purpose he first used the nitrate of silver, which was mentioned to him by a friend as a substance very sensible to the influence of light; but all his numerous experiments, as to their primary end, proved unsuccessful."

It will be seen, from the foregoing account of the results of their experiments, that Mr. Wedgwood's process and the early processes of Mr. Talbot were nearly alike; and if he had possessed the means which the compound salt hyposulphite of soda afforded to subsequent photographers, of destroying the sensibility of the prepared paper to further impressions of the rays of light, there can be little doubt that the invention would have attained a high degree of perfection at the commencement of the present century. As it was, the failure of Mr. Wedgwood to accomplish the object he was so nearly attaining appears to have discouraged attempts by others, and twenty years elapsed without any advance having been made towards its realization.

M. Niepce, of Chalons on the Saone, who was the first to succeed in obtaining permanent representations of the images of the camera, commenced experimenting on the subject in 1814, at least ten years before

M. Daguerre directed his attention to Photography. In 1826 these two gentlemen became acquainted, and conjointly prosecuted the investigations which led to the beautiful result of the Daguerreotype. M. Niepce having previously succeeded in obtaining durable representations of the pictures focused in the camera, he came to this country in 1827, and exhibited several of the results of his process, and communicated to the Royal Society an account of his experiments. These photographs, which may be considered the first durable ones that had been obtained, were, with one exception, taken on plates made of pewter. One of the largest was 5¼ inches long and 4 inches wide. It was taken from a print 2½ feet in length, representing the ruins of an abbey. When seen in a proper light, the impression appeared very distinct. Another one, which was stated to have been the first successful attempt, was a view taken from nature, representing a court-yard. Its size was 7½ inches by 6 inches, but it was not so distinct as the preceding one. A third specimen was an impression on paper, *printed from a photograph on metal*, the picture having been etched into the plate by nitric acid, and then printed from. All these specimens, though extremely curious as the first successful attempts to preserve the images of the camera, were more or less imperfect, and were far from presenting the beautiful results of Photography now attained. It is remarkable, however, that the original process of etching the picture on a metal plate, and printing from it, has now, in the perfected state of the art, become the most recent improvement;

and the prints from photographic plates present some of the most beautiful effects hitherto produced.*

M. Niepce communicated the particulars of his process to M. Daguerre in December, 1829. They then entered into an agreement to pursue their investigations jointly, but it was not until ten years afterwards that the invention of the Daguerreotype by M. Daugerre was made known. To M. Niepce must, therefore, be awarded the honour of having first discovered the means of rendering permanent the transient images of the camera obscura. The plan he adopted was to cover a plate of white metal with asphalte varnish, and expose it to the action of light in a camera, when the parts whereon the light was concentrated became hardened, and the other parts remained unaltered, and could be washed away.

In M. Niepce's account of the process, after describing the preparation of the asphalte varnish, he says:—"A tablet of *plated silver*, or well-cleaned and warm *glass*, is to be highly polished, on which a thin coating of varnish is to be applied cold, with a light roll of very soft skin. This will impart to it a fine vermilion colour, and cover it with a very thin and equal coating. The plate is then placed on heated iron, which is wrapped round with several folds of paper, from which, by this method, all moisture has been previously expelled. When the varnish has ceased to simmer, the plate is withdrawn from the heat

* The original photographs produced by M. Niepce are still preserved in good condition, and were last year exhibited at the Royal Institution.

and left to cool and dry in a gentle temperature, and protected from a damp atmosphere. The plate, thus prepared, may be immediately subjected to the action of the luminous fluid in the focus of the camera; but even after having been thus exposed a length of time sufficient for receiving the impressions of external objects, nothing is apparent to show that these impressions exist. The forms of the future picture remain still invisible. The next operation then is to disengage the shrouded image, and this is accomplished by a solvent."

The solvent employed was a mixture of one part of oil of lavender, and ten parts of oil of petroleum. The solvent was poured over the plate, and allowed to remain. M. Niepce continues: "The operator, observing it by reflected light, begins to perceive the images of the objects to which it has been exposed gradually unfolding their forms, though still veiled by the supernatant fluid, continually becoming darker from saturation with the varnish."

The time required for the exposure of the plates in the camera was six or eight hours. For the purpose of darkening the pictures, M. Niepce used iodine, and it has been supposed that the use of iodine for that purpose suggested the employment of it to his partner.

The process adopted by M. Daguerre was, to deposit a film of iodine on a highly polished silver plate, by exposing the plate to the vapour of iodine in a dark box. The prepared plate was then placed in the camera, and after an exposure of ten minutes or more, according to the brightness of the day, an impression was made on the iodised silver, but too faint to be visible. To bring out the image thus invisibly impressed, the plate was exposed to the vapour of mercury, in a closed box. The mercury adhered to the parts on which the light had acted, and left the other parts of the plate untouched; and by this means a beautiful representation was produced, in which the deposited mercury represented the lights of the picture, and the polished silver the shadows. The iodised silver remaining on the plate not acted on by light, was washed away by a solution of hyposulphite of soda, and the picture could then be exposed without injury.

Nothing can exceed the delicacy of delineation by such a Daguerreotype; for the fine surface of the highly polished silver seems to exhibit the impressions of the smallest objects that emit rays of light. The length of time required to produce an impression was, however, a serious obstacle to the use of the process, as originally invented, for taking portraits. Numerous attempts were consequently made to obtain a more sensitive material. Bromine was tried, in addition to iodine, and with such complete success, that a few seconds were sufficient to effect an impression on the plate, which could be forcibly brought out by the vapour of mercury.

It was in 1840 that portraits were first taken by the Daguerreotype process in this country. In the first instance, a concave mirror was employed to concentrate the rays of light on the plate, instead of a lens; and the author has now in his possession a portrait taken in this manner, by "Wolcott's reflecting apparatus." The object of using a concave mirror was to be able to concentrate a greater number of the rays of light than could be done by a lens, and thus to form a brighter image. At the time that portrait was taken, the means had not been discovered of making the mercury adhere to the plate, and a feather would brush it away. Soon afterwards, however, M. Fizeau ingeniously contrived to fix the images on the plate by gilding it. This was done by pouring on to the plate a few drops of a diluted solution of muriate of gold, and holding it horizontally over the flame of a spirit lamp; by which means the gold was deposited and formed a thin, beautiful film of the metal over the surface, and thus not only made the picture more durable, but gave it increased effect.

The French government, fully appreciating the importance of the invention, determined to purchase it from the patentee, and to throw it open to the public. An account of the invention was published in June, 1839; and in the following month an arrangement was entered into, to the effect that, in consideration of M. Daguerre making the process fully known, a pension of 6,000 francs should be granted to him for life, and a pension of 4,000 francs to M. Isidore Niepce, the nephew of the original inventor of Photography, his uncle having died before the final success was attained.

It was generally supposed at the time, that by the grant of those pensions the invention was thrown open to the whole world, as represented by the French

Minister; but, nevertheless, M. Daguerre patented the process in other countries, and France alone reaped the benefit of a free use of the invention.

Whilst M. Daguerre was thus successfully working out to perfection the plan of producing beautiful naturally-impressed pictures on iodised silver surfaces, Mr. Fox Talbot was at the same time nearly attaining the same results. The following is the account given by himself of his researches:*—"Having in the year 1834 discovered the principles of Photography on paper, I some time afterwards made some experiments on metal plates; and in 1838 I discovered a method of rendering a silver plate sensitive to light, by exposing it to iodine vapours. I was at that time, therefore, treading in the footsteps of M. Daguerre, without knowing that he, or indeed any other person, was pursuing, or had commenced or thought of, the art which we now call Photography. But as I was not aware of the power of mercurial vapour to bring out the latent impressions, I found my plates of iodised silver deficient in sensibility, and therefore continued to use in preference my photogenic drawing paper. This was in 1838. Some time after—in August, 1839—M. Daguerre published an account of his perfected process, which reached us during the meeting of the British Association; and I took the opportunity to lay before the Section the facts which I had myself ascertained in Metallic Photography."

Whilst to M. Daguerre must be awarded the hon-

* "Philosophical Magazine," February, 1843.

our of having first brought to perfection the method of rendering permanent the images of the camera on metal plates, Mr. Fox Talbot may claim to be the first who perfected similar images on paper, which the comparative roughness of the surface alone prevented from being as delicately beautiful as the pictures of the Daguerreotype. He commenced his experiments in Photography in 1834; and on the 31st of January, 1839, he read a paper before the Royal Society, entitled, "Some Account of the Art of Photogenic Drawing; or, a process by which natural objects may be made to delineate themselves without the aid of the artist's pencil."

Mr. Talbot had not then succeeded in obtaining the impressions of images focused in the camera; what he had succeeded in doing was to fix upon paper the shadows of objects placed upon it, and exposed to the light of the sun. The paper was first dipped into a solution of common salt, and then wiped dry, to diffuse the salt uniformly through the substance of the paper. A solution of nitrate of silver was then spread over one surface with a soft brush, and dried carefully before a fire in a darkened room. The strength of the solution was regulated by first obtaining a saturated solution of the nitrate of silver, and afterwards diluting it with six or eight times its volume of water. The objects to be copied, such as leaves, lace, or other flat surfaces, were pressed against the prepared paper by a glass fixed in a frame, and exposure to light quickly darkened all the parts of the paper, excepting those shaded by the objects. The image thus impressed was what is termed a "negative," the dark parts which excluded the light being left white on the paper, and the parts through which the light passed being darkened. To produce a picture corresponding with the natural lights and shades, the process was repeated, substituting the picture first obtained, on thin transparent paper, for the original object, by which means the lights and shadows were reversed.

The chloride of silver, formed on the surface of the sensitive paper by the combination of the common salt and nitrate of silver, being insoluble in water, great difficulty was experienced in washing it away, so as to prevent the whole surface from afterwards darkening on exposure to light. The application of hyposulphite of soda, for the purpose of making the pictures durable, was suggested by Sir John Herschel, and it answers remarkably well, as it dissolves the chloride of silver. A solution of ammonia is nearly equally efficacious in removing the chloride.

The Calotype process, by which the images of the camera can be fixed upon paper, was invented by Mr. Talbot, in 1840. It is thus described:—Dissolve 100 grains of crystallized nitrate of silver in 6 ounces of distilled water. Procure some fine writing paper, and wash one side of it with the solution, laid on with a soft brush; then dry the paper cautiously, by holding it at a distance from the fire. When dry, dip the paper into a solution of iodide of potassium, containing 500 grains dissolved in 1 pint of water, and let it remain in the solution two or three minutes. Then dip it into a vessel of water; remove the water on the surface by blotting paper, and dry it by a fire, in the dark or by candle-light. The paper thus prepared is called "iodised paper;" it is not very sensitive to light, and may be kept for some time without spoiling. Next dissolve 100 grains of crystallized nitrate of silver in 2 ounces of distilled water; add to the solution one-sixth of its volume of strong acetic acid, and call that mixture A. Then make a strong solution of crystallized gallic acid in cold water, and let that solution be called B. Mix equal volumes of A and B together in small quantities at a time. That mixture Mr. Talbot calls gallo-nitrate of silver, and with it wash over the surface of the iodised paper. Allow the paper to remain half a minute, and then dip it into water, and again dry it lightly with blotting paper. The paper thus prepared is very sensitive, and will receive an impression in the camera in the shortest possible time. The impression is at first invisible, but it may be brought out by laying the paper aside in the dark, or by washing it once more in the gallo-nitrate of silver, and holding it at a short distance from the fire. To fix the picture, the paper is first washed in water and lightly dried, and then soaked in a solution of hyposulphite of soda for a few minutes, by which means the iodised silver is removed, and after being again washed in water and dried, the process is completed. The picture thus produced is a negative one, and requires to be transferred in the manner before stated. The original Calotype may, by that means, serve to produce a great number of pictures.

Mr. Talbot's patent was sealed on the 8th of

February, 1841. In his specification, he claimed the use of gallic acid, and he succeeded in enforcing his claim in a Court of Law, though it appeared that on the 10th of April, 1839, photographs of objects taken in the solar microscope in five minutes, by the Rev. J. B. Reade, were shown at the London institution, which were described to have been produced by an infusion of galls, and fixed with hyposulphite of soda. It must be mentioned, however, to Mr. Talbot's honour, that on a representation to him by the President of the Royal Society that the art of Photography was impeded in its progress in this country by patent monopolies, he generously made a present to the public of all his inventions and discoveries, reserving to himself only the privilege of taking portraits.

The transfer from one paper to another of the picture obtained in the camera, and the comparative roughness of the surface of the paper itself, prevent Calotypes from exhibiting that sharpness and delicacy of definition which are so admirable in a Daguerreotype. Several attempts were therefore made to obtain a more smooth surface for the reception of the image; but without much success, until glass was adopted for the purpose. To make that material available, it is necessary to coat it with some substance that will absorb the sensitive solution. In the first instance, the white of eggs was employed with considerable success. Albumen has, however, been supplanted by collodion—a solution of gun-cotton in ether—which is found to be peculiarly suitable for the reception of the sensitive preparation of silver.

In conducting the collodion process, the collodion is first iodised by adding to it iodide of potassium and iodide of silver, dissolved in alcohol. The iodised collodion is then poured over a plate of glass that has been carefully cleaned, and is moved about horizontally until a perfectly uniform film is spread over the surface, to which it adheres firmly. The plate is afterwards dipped into a solution of nitrate of silver, which renders it so highly sensitive to impressions of light, that it will receive an image in less than a second. The image is latent, until it is developed by pouring over the plate a mixture of pyro-gallic acid in distilled water, acetic acid, and nitrate of silver. The impression is fixed with hyposulphite of soda.

The pictures produced by the collodion process are negatives, which serve admirably for transferring positive pictures on to sensitive paper. But, if required, the negative picture can be readily changed into a positive one, by converting the darkened silver into white metallic silver, by a mixture of protosulphate of iron and pyro-gallic acid. In a short time a white metallic image is obtained, which, when relieved by a background of black velvet or black varnish, equals in delicacy of finish the most beautiful Daguerreotypes.

Many attempts have been made, but hitherto without success, to obtain photographs coloured, as well as shaded, by nature. The opinions of those who have most studied the subject differ as to the possibility of ever attaining that desired object. Sir John Herschel has so far shown that it is not impossible, as to have impressed the colours of the solar spectrum on paper, by the mere action of light; and parts of the images of objects fixed on the screen of the camera are also sometimes coloured. These facts induce us to hope that in the progress of discovery some means may be found of obtaining naturally-coloured photographs, notwithstanding it has been pronounced, by good authorities, to be an absolute impossibility.

Specimens of coloured photographs were exhibited by Mr. Mercer at the recent meeting of the British Association, which showed that by the use of various chemical preparations that are sensitive to light, photographs may be shaded in colours. The principal re-agents employed were salts of iron, and by immersing the paper in suitable menstrua, after the image had been impressed in the camera, the picture was developed in any colour required; the same tint being spread over the whole. One purpose to which it was suggested this coloured photographic process is applicable, is printing on woven fabrics, the action of light serving as a mordant to fix the colours.

Photography has been already applied to various uses, and it is capable of being rendered much more valuable. To the meteorologist it affords the means of registering the rise and fall of the mercury in the barometer and thermometer, and, by a self-registering apparatus, the changes of temperature and of atmospheric pressure are marked upon paper that records the time at which the changes occur. It may also be applied, in the same manner, to register the directions of the wind, and the times of its changes. The sun impresses his own image upon paper; and the spots on his surface, thus correctly delineated, can be compared with those seen in pictures of the sun at other times; and the foundation is laid for more correct knowledge of the nature of those appearances, and of the motion of the sun himself. Photographs of the moon and planets present exact representations of those heavenly bodies, as seen through the most powerful telescope; and, with the assistance of the stereoscope, the figure of the moon is shown in its true globular form, as it can be seen by no other means. It has been proposed, indeed, by the aid of Photography, to extend our knowledge of the stars far beyond the reach of telescopic vision; for as the image focused on the screen of the camera is composed of rays from every object on the body of a star, it might be possible to see those objects by greatly magnifying the image. It remains, however, for the further progress of discovery and invention, to arrive at so delicate a delineation by photographic processes, as to obtain landscapes of the moon, and portraits of the inhabitants of Jupiter!

One of the latest advances in the art of Photography has been the engraving on steel-plates by the action of light, by which means more forcible effects have been obtained than by the impressions of light upon paper. Mr. Fox Talbot has distinguished himself in thus fixing the images on steel, as he was the first to impress them upon paper. In his method of doing so, he covers the steel plate with a solution of isinglass and bichromate of potass, and placing a collodion negative picture upon it, he exposes it to the action of light. When the picture is sufficiently impressed, he etches it into the plate by means of bichloride of platinum. M. Niepce, the nephew of the original inventor of Photography, has produced the same effect by reviving the first processes adopted by his uncle; using, as he did, bitumen, dissolved in essential oil of lavender, to cover the plates. Two other foreign photographers, M. Poitevin and M. Pretschi, have also successfully directed their attention to engraving the images of the camera, which has now obtained a high degree of perfection.

It is well worth notice that these most recent improvements in Photography are but further developments of the original designs of M. Niepce, who not only succeeded in etching the pictures impressed by the light of the sun on his metal tablets, but made use of a glass surface, on which the now generally adopted collodion process depends.

DISSOLVING VIEWS.

There are no optical illusions more extraordinary than those shown in the exhibition of Dissolving Views. The effects of the changes in the diorama are only such as are seen in nature, the same scene being represented under different circumstances, and the marvel in that case is that such beautiful and natural effects can be produced on the same canvas. But Dissolving Views set nature at defiance, and exhibit metamorphoses as great as can be conceived by the wildest fancy.

Whilst, for instance, the spectator is looking at the interior of a church, he sees the objects gradually assuming different appearances. The columns that support the vaulted roof begin to fade away, and their places are occupied by other forms, which gradually become better defined and stronger, and a tree, a house, or, it may be, a rock, thrusts the columns out of view, and the roof dims into blue sky, chequered with clouds. The original view thus entirely disappears, and the scene is changed from the interior of a church to open country, or to a rocky valley. This is done, not by changing at once one scene into another, but by substituting different individual objects, which at first appear like faint shadows, and then, becoming more and more vivid, at length altogether supplant their predecessors on the field of view, and will, in their turn, be extinguished by others.

It sometimes happens that some strongly marked object resists apparently the efforts made to dispossess it, and in the midst of a mountainous scene will be observed the form of a chandelier or of a statue, that occupied a distinguished place in the church that has just vanished. In a short time, however, these relics disappear, and the mountain, the valley, and the lake are freed from the incongruous images of the former scene.

These effects are produced in a manner as simple as they are extraordinary. All that is requisite is to have two magic lanterns fitted on to a stand, with their tubes inclined towards each other, so that both discs of light may exactly coincide, and form on the screen a single disc. If paintings on glass, representing different views, be then placed in each lantern, with the lenses adjusted to bring the rays to a focus on the screen, the two images will be so mingled together as to present only a confused mixture of colours. Suppose one of the views to be the interior of a church, and the other to be a mountain scene ;—the pillars of the church will be mingled with trees and rocks, and in the midst of the confusion there may perhaps be discerned a strongly painted chandelier or an altar

piece. When an opaque shade is placed before the lens of either of the lanterns, to prevent the light from reaching the screen, the previous confusion becomes instantly clear and distinct, and the church or the landscape is seen without any interfering images. If the opaque screen be gradually withdrawn from one lens, and at the same time drawn in an equal degree over the other, the different objects will again be mingled, and those in the one scene will predominate over those in the other in proportion to the rela-

tive quantities of light permitted to issue from each lantern to the screen. The two first of the accompanying drawings are thus blended together in the third, when the screen is half withdrawn from each.

It is usual to fix the opaque shade, which alternately covers and exposes the two magic lanterns, on to a central pin, so that it may be moved vertically up or down. The shade is so arranged, that in raising the end to cover the lens of one lantern, the farther end descends, and exposes, in an equal degree, the other lens. During the time that either of the views is altogether concealed, the painting is changed; and in this manner an unlimited number of metamorphoses may be effected.

It requires no expensive apparatus to show the effect of Dissolving Views on a small scale. Two common magic lanterns are quite sufficient for the purpose of private exhibition, and the angle at which they should be fixed on their stand may be readily ascertained after a few trials. To make the transformation more extraordinary, a man's face may be painted on one glass and a landscape on the other; and, when the change is made from the face to the landscape, a strongly painted eye or nose may be seen occupying the centre of the view, long after the other features have disappeared, until all the rays of light from that painting have been excluded. The change from youth to age, from beauty to ugliness, may also be shown with striking effect.

It will be observed that the principle, on which the metamorphoses of Dissolving Views depend, is similar to that which produces the variations in the diorama. In both cases there are two paintings on the same space, either of which may be shown at pleasure by different dispositions of the light; the chief difference between them being that the Dissolving Views are seen altogether by reflected light, whilst in the diorama the paintings at the back and front are shown alternately by reflected and by transmitted light.

THE KALEIDOSCOPE.

No invention, on being first brought out, created so general a sensation as the Kaleidoscope. Every person, who could buy or make one, had a Kaleidoscope. Men, women, and children—rich and poor; in houses or walking in the streets; in carriages, or on coaches—were to be seen looking into the wonder-working tube, admiring the beautiful patterns it produced, and the magical changes which the least movement of the glass occasioned.

It was in the year 1814 that Sir David Brewster discovered the principle on which the effects of the Kaleidoscope depend, whilst he was engaged in experiments on the polarization of light by successive reflections between plates of glass. The reflectors were in some cases inclined to each other, and he remarked the circular arrangement of the images of a candle round a centre. In afterwards repeating the experiments of M. Biot on the action of fluids on light, he placed the fluids in a trough formed by two plates of glass cemented together at an angle. The eye being placed at one end, some of the cement which pressed through between the plates appeared to be arranged in a circular figure. The symmetry of this figure being very remarkable, Sir David Brewster undertook to investigate the cause of the phenomenon, and the result of his investigations was the invention of the instrument to which he gave the name of Kaleidoscope, from the Greek words *καλος*, beautiful, *ειδος*, a form, and *σκοπεω*, to see. *

The Kaleidoscope in its simplest form consists of two equal strips of plate glass, about 8 inches long and 2 inches wide, silvered on one side, to act as reflectors. These glasses are placed one over the other exactly, and then the edges on one side being separated, whilst the two other edges are kept close tegether, they are fixed by means of separating pieces of wood and string at the angle required. The glasses are then fitted into a metal tube, which has an eyehole at one end, and at the other end of the tube there is fixed a small cell of ground glass, to contain pieces of differently stained glass or other objects, that are to be multiplied by reflections into beautiful symmetrical figures. In the better kind of Kaleidoscopes, the cell containing the objects may be turned round, by which means the pieces of glass shift their positions, and the figures instantly change. The same effect is produced, though in a less agreeable manner, in the common kind of instruments, by turning the tube.

To form by the combined reflections from the two glasses a perfectly symmetrical figure, the sector comprised between the inclined sides of the glasses may consist of any even aliquot part of a circle. In

* Brewster's Encyclopædia, article "Kaleidoscope."

the accompanying diagram, the ends of the flat silvered glasses *a c*, *b c*, are inclined at an angle of 60 degrees; therefore the circle is completed by the junction of six sectors. In such a Kaleidoscope, the circular figure will be formed by three reflections from each glass

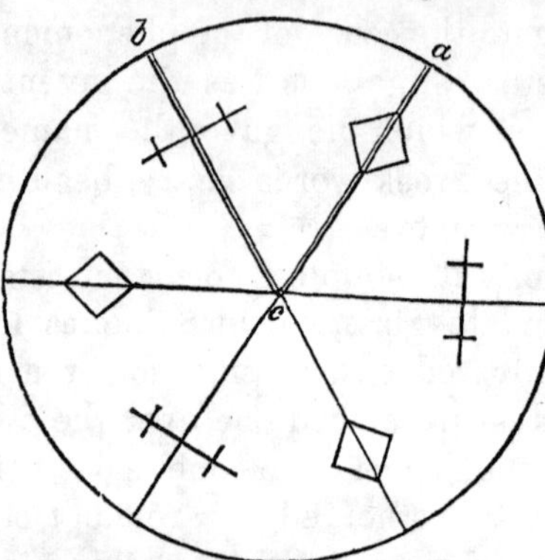

To make the formation of the circular figure by repeated reflections more intelligible, we will consider it as composed of the smallest possible number of equal divisions, as in the second diagram, in which the circle is divided into quadrants. In such an arrangement of the reflectors, the figure seen on looking through the central aperture will consist of four parts. In the first place, the objects included in the space *a b c*, between the inclined glasses, will be seen directly by rays of light from the objects themselves; viz., the small cross *d*, and the triangle *e*. The same field of

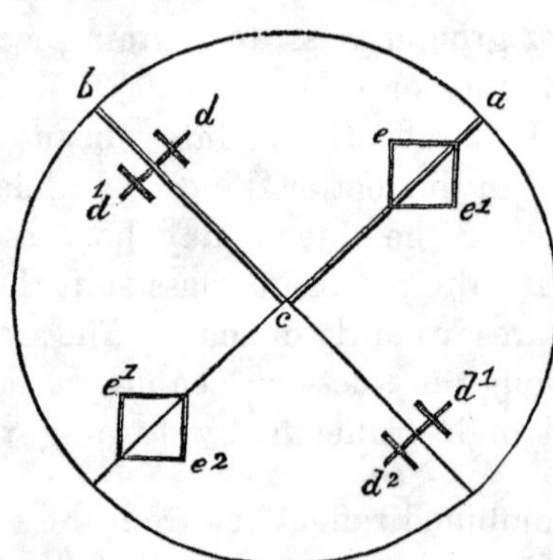

view will be reflected from both mirrors, by which reflection the cross on one side will seem to be doubled, and the triangle on the other will have another similar one added to it, to make a complete rhomb. The cross will also be reflected by the mirror on the right side, and the triangle by the one on the left. The images of the objects contained within the space *a b c*, being thus presented by reflection on both sides, they become the objects for further reflections from parts of the mirrors still nearer the spectator. Thus the images d^1 on both sides are reflected to form the single image d,^{2} and the images e^1 are in the same manner reflected to form the second image e^2.

When the angle formed by the inclination of the mirrors divides the circle into a greater number of sectors, the reflections of the images are repeated, from points nearer and nearer to the eye, and the circle is thus completed, however numerous the sectors may be; but at each repetition of the reflection, the images will become more dim, since, owing to the imperfection of reflecting surfaces, a portion of the light is absorbed at each reflection.

In the first instruments that were constructed, the objects were fixed in the field of view, therefore scarcely any change of pattern was obtainable. It was not until some time afterwards that the idea occurred to Sir David Brewster of producing endless changes of the figures, by making the objects movable in a cell of glass at the end of the instrument. He afterwards introduced other improvements in the Kaleidoscope, for extending its range of objects, for varying the angles of inclination, and for projecting the figures on a screen. In the instrument, as ordinarily made, the objects to be seen properly must be placed close to the end of the reflectors; but by the addition to the instrument of a tube containing a lens, the rays from distant objects are brought to a focus near the mirrors, and the image formed there is repeated by the reflectors in the same manner as a solid object.

The projection of the figures on a screen, by an apparatus similar to a magic lantern, gives great additional pleasure to the effects of the Kaleidoscope, as the figures are not only seen by several persons at the same time, but they are presented in a magnified form. The projection of the figures also increases the use of the instrument in designing patterns, for which purpose it has been employed with great advantage.

A patent for the Kaleidoscope was taken out in 1817, but the high prices charged by the opticians who were authorized by the inventor to sell the instrument, and the facility with which it could be made, occasioned a general violation of the patent right, and it was not long before the claim of Sir David Brewster, as the original inventor, was disputed. In the indignant vindication of his claim, he observes:—"There never was a popular invention which the labours of envious individuals did not attempt to trace to some remote period;" and the Kaleidoscope was not an exception. It was found that Kircher had described the effects of repeated reflections as far back as 1630; and that Mr. Bradley had, in 1717, made a philosophical toy, consisting of two small mirrors, that opened like a book, which, when partially opened, repeated the reflections of objects placed near it in the same manner as the Kaleidoscope. But this instrument was so different in its construction, and in the effects it produced, from the Kaleidoscope, that Sir David Brewster's claim to be the inventor may be freely admitted. The fact that it took the world by surprise, and created a sensation greater than any other invention had done before, is sufficient to establish its title as an original invention.

THE MAGIC DISC.

There are several ways of illustrating the retention by the retina of the eye of the images of objects after they have been withdrawn from sight, but none is so curious as the philosophical toy called the Magic Disc, which, from the optical principles involved in its extraordinary effects, deserves to be noticed as one of the remarkable inventions of the present century.

One of the most striking methods of exhibiting the retentive property of the retina, before the invention of the Magic Disc, was to paint different objects at the back and on the front of a card, and by then giving rapid rotation to the card, both objects were seen together. Thus, when the figure of a bird is painted on one side, and an empty cage on the other, by rapidly turning the card, the bird appears to be in the cage. In the Magic Disc the objects are painted on the same side of a circular piece of card-board, and both are exposed to view during their rapid rotation.

The disc is divided into eight or ten compartments, in each one of which the same figures are repeated, though the positions of one or more of them are changed. A favourite subject represented is a clown leaping over the back of a pantaloon, which affords a simple illustration of the apparent relative movements of two bodies, and will serve to explain how the effect is produced.

The instrument consists of a disc of stiff card-board, about nine inches diameter, mounted on a horizontal pivot in the centre, on which it may be freely turned. Between each of the compartments of the disc there

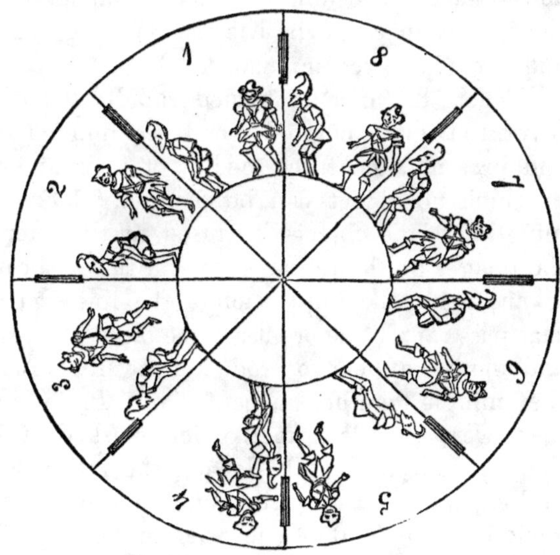

is an elongated aperture, about one inch long and a quarter of an inch wide, for the eye to look through. Suppose the disc to be divided into eight compartments, by radial lines. In the compartment No. 1, the pantaloon is represented in a stooping posture, and the clown is on the ground ready to make a spring. In No. 2 the pantaloon is in the same attitude, but the clown has commenced his leap, and is raised a little way from the ground. In the third division he is shown still higher in the air; and in the fourth he is mounted above the shoulders of pantaloon, who retains the same posture as at first. The fifth compartment represents the clown as having jumped over pantaloon's head, and coming down to the ground; and in each succeeding division his farther descent is shown, till, in No. 8, he has reached the ground again, and is ready to recommence the leap.

When the disc is turned rapidly round on its pivot, the figures painted upon it are mingled together, and present a confused medley of lines and colours, in which no object can be distinctly defined. This mingling of the objects is caused by the retention of the images by the retina, so that if the eye be directed to any point, the impression of the lines and colours that pass rapidly before it is not effaced before another and another appear to produce fresh impressions, and they mingle together in confusion. If, for instance, there were a circle formed of dots marked on the disc, the impression of each dot on the retina would be prolonged; and as, by the rotation, other dots would come into the field of view before the impression of the first was removed, it would form an unbroken ring. But if the disc were screened from sight, at intervals of nearly equal duration to that of the continuous impression, so as to efface the image of one dot before the rays of another were admitted to the eye, then the ring would be seen to be composed of dots, as distinctly as when the disc was stationary.

The effect of screening the objects from the eye at short intervals is produced by looking with one eye through the openings at the image of the disc, reflected from a mirror. The figures are then seen only when the apertures come opposite the eye; but as the impression of one view remains till it is renewed by the light admitted through the next aperture, there is continuous vision of the objects painted on the disc.

It is thus that the figures of pantaloon and clown become visible, and their apparent relative movements are occasioned. For instance; each time that the impression of the figure of the pantaloon is renewed, he is seen in the same place and in the same attitude; therefore he appears to be stationary, though the successive pictures that compose his figure to the eye are in rapid rotary motion. The figure of the clown, however, is seen in a different position each time that he comes into view, therefore he appears to be in motion relatively to pantaloon, though stationary as regards his absolute position on the disc.

The same effect would be produced if the disc, during its rotation, were seen by successive electric sparks. The electric spark is so momentary in its duration, that the most rapidly moving objects appear stationary; therefore each spark would show a seemingly stationary disc, on which the figure of the clown would appear in different relative positions; and the illusion would be as perfect as when the rays of light are interrupted at intervals.

The electric spark is so instantaneous that a cannon ball might be seen in its rapid flight, if illuminated by a flash of lightning, and would seem to be stationary. Professor Faraday mentioned, in one of his lectures, the extraordinary appearance which a man, who was jumping over a stile, presented when seen by lightning on a dark night. The man seemed to be resting horizontally in the air, with one hand touching the stile.

The duration of the impression of an object on the retina is capable of illustration by means of the Magic Disc in a great variety of designs, each one of which may represent many movements. The turning of the wheels of machinery, the tossing of balls, the dancing figures of men and women may thus be shown, the designs for which afford ample scope for exercising the pencil of an ingenious artist.

THE DIORAMA.

Those who are old enough to remember the Regent's Park before there were any houses northward of the New Road, may recollect that among the first buildings erected, on what is now called Park Square, was a strange-looking, partly semi-circular erection, provided with ample lighting space, which attracted great attention during its progress, and was the cause of much speculation as to its probable purpose. That building was intended for the exhibition of the Diorama.

M. Daguerre, the inventor of the Daguerreotype, had, in conjunction with M. Bouton, a short time previously opened a similar exhibition in Paris, where the beauty of the paintings, aided by the extraordinary effects of newly contrived dispositions of the light, had excited a great sensation. The Diorama was opened in London on the 6th of October, 1823, and for a long time it was equally popular in this metropolis.

The visitors, after passing through a gloomy anteroom, were ushered into a circular chamber, apparently quite dark. One or two small shrouded lamps placed on the floor served dimly to light the way to a few descending steps, and the voice of an invisible guide gave directions to walk forward. The eye soon became sufficiently accustomed to the darkness to distinguish the objects around, and to perceive that there were several persons seated on benches opposite an open space, resembling a large window. Through the window was seen the interior of a cathedral, undergoing partial repair, with the figures of two or three workmen resting from their labour. The pillars, the arches, the stone floor and steps, stained with damp, and the planks of wood strewn on the ground, all seemed to stand out in bold relief, so solidly as not to admit a doubt of their substantiality, whilst the floor extended to the distant pillars, temptingly inviting the tread of exploring footsteps. Few could be persuaded that what they saw was a mere painting on a flat surface. This impression was strengthened by perceiving the light and shadows change, as if clouds were passing over the sun, the rays of which occasionally shone through the painted windows, casting coloured shadows on the floor. Then shortly the brightness would disappear, and the former gloom again obscure the objects that had been momentarily illuminated. The illusion was rendered more perfect by the excellence of the painting, and by the sensitive condition of the eye in the darkness of the surrounding chamber. Whilst gazing in wrapt admiration at the architectural beauties of the cathedral, the spectator's attention was disturbed by sounds underground. He became conscious that the scene before him was slowly moving away, and he obtained a glimpse of another and very different prospect, which gradually advanced until it was completely developed, and the cathedral had disappeared. What he now saw was a valley, surrounded by high mountains capped with snow. This mountain valley seemed scarcely less real than the arched roof and columns of the cathedral, whilst a foaming cascade, dashing down the rocks, and the sound of rushing waters, added to the illusion. After looking for some time at this beautiful valley, the clouds were seen to gather on the mountain tops, and a storm impended. A gleam of sun-light, still resting on the edge of the clouds, exhibited a strange contrast between the silvery brightness and the dense black vapour that shrouded the hills, and could almost be felt. It was but a passing thunderstorm. Presently the dark clouds rose from the valley, and dispersed; the sun again shone on cottage, vineyard, and mountain, charming the spectator as much by the beauty of the scene as he was astonished by the wonderful change.

Such was the Diorama as it was first exhibited in London to admiring crowds. In subsequent years greater changes were made in the variations of light and shade; and by the introduction of mechanical contrivances, with more or less success, the magical effects were increased, without, however, adding to the apparent reality of the objects. A church or cathedral was always the subject of one view, and sometimes of both. The interior of an empty church would be shown by evening twilight. The shades of evening gradually darkened into the obscurity of night, and then the glimmer of candles would be seen spreading more and more widely, until the church was lighted up, and it was occupied by a crowded congregation at midnight mass. Some views represented the exterior of a ruin or of a cathedral after sunset, and as night advanced, the stars twinkled in the blue sky, and the moon rose and threw its silvery light on water, buildings, and clouds, contrasting in some cases with the red glare of lamps from the windows of houses and shops. The disc of the moon exactly resembled that of the real luminary, and all around being so dark, the rays from its surface cast shadows of intervening objects. In one picture a still more astonishing appearance was produced, by the change of the interior of a beautifully painted and decorated church into a mass of charred ruins.

The means principally adopted for the production of these magical changes in a painting on a flat surface, and for giving such seeming reality to the objects represented, were for some time kept secret; nor do we think they are even yet much known. As in many other clever inventions, the effects are produced in a very simple manner. The picture is painted on both sides of a transparent screen, and the change of scene is occasioned almost entirely by exhibiting the picture at one time by reflected light, from the surface nearest the spectator, and afterwards by transmitted light, after excluding the light from the front.

Let us take for illustration the interior of a church,

at first seen empty, and afterwards filled with people, and illuminated by candles. The empty church is painted on the front on fine canvas or silk, in transparent colours, and at the back are the figures and candles, and other objects intended to appear with them. The arrangements for illuminating the picture are so contrived, that the light may be thrown entirely on the front or on the back, or partly on both. When the light is on the front, the empty church only is visible. It is then gradually darkened, and the back of the picture is illuminated, by which means the figures and candles are seen; and the form of the building being preserved, the same church, which was before empty, becomes occupied by a crowded congregation.

It may be mentioned, as an illustration of the perfect illusion of the Diorama, that a lady who on one occasion accompanied the author to the exhibition, was so fully convinced that the church represented was real, that she asked to be conducted down the steps to walk in the building.

The effect of changing the direction of the light may be readily perceived by making a drawing on both sides of a sheet of paper, as shown in the annexed engraving. The side backing this page represents the interior of St. Paul's Cathedral when empty, and on the back several figures are drawn. Those figures are invisible until the leaf is held up against the light, and when the drawing is seen as a transparency, the objects on the back, as well as those in front, come into view, and the building appears to be occupied.

Any one who has a taste for drawing, and a little ingenuity, may thus produce many pleasing and astonishing effects. It will be desirable to procure, in

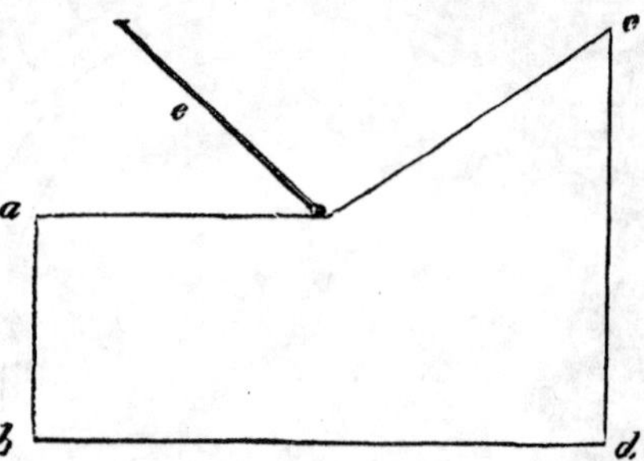

the first instance, a box, so contrived that it will hold the painting, and afford the means of throwing the light on the front or on the back at pleasure. The diagram shows the form of such a box. The letters *a*, *b*, *c*, *d* mark the outside; the aperture, at *c d*, being enlarged to permit several persons to look into it at the same time. The box may be of any required dimensions, to suit the size of the drawing, which is to be fitted into a groove at *a b*, and the interior must be blackened. The lid, *e*, when open, as in the diagram, admits the light to the front of the picture, the back being covered with an opaque screen. As the lid is closed, the picture becomes darkened, and by the gradual removal of the screen at the same time, it is changed into a transparency. This portable Diorama can be most conveniently shown by lamplight, the flame of an argand lamp, the wick of which can be heightened and lowered, being best adapted for the purpose. The effect by daylight is, however, superior, but the room must then be darkened, and the admission of light confined to the picture.

The moving water, and the motion of smoke and clouds, which were frequently introduced in the Diorama, were mechanical additions, the effects being produced by giving motion to bodies behind, the forms of which were seen by transmitted light. The introduction of such mechanical aids, however, detract from the artistic character of the Diorama, the principal merit of which consists in exhibiting the changes occasioned by variations in the mode of throwing the light on the two-faced picture.

It is to be regretted that exhibitions of a larger and more showy kind should have superseded the Diorama in public estimation; and that, from the want of support, their charming and marvellous pictorial representations, which formed, in days gone by, one of the principal "sights" of London, should be now closed.

THE STEREOSCOPE.

One of the most beautiful as well as the most remarkable pictorial illusions is produced by the combination of two views into one by the recently invented instrument called the Stereoscope. In the Diorama, in the Magic Disc, and in the Dissolving Views, separate paintings combine to produce different effects; but in the Stereoscope the two pictures unite into one to give additional effect to the same view, and to make that which is a flat surface, when seen singly, appear to project like a solid body.

The principle of the Stereoscope depends on the different appearance which near objects present when seen by the right or by the left eye. For instance, on looking at a book placed edgewise, with the right eye, the back and one side of the book will be perceived; and on closing the right eye and opening the left, the back and the other side of the book will be seen, and the right-hand side will be invisible. It is the combination of both these views by vision with two eyes that produces the impression of solidity of objects on the mind; and if the different appearances which the book presents to each eye be copied in separate drawings, and they can afterwards be placed in such a position as to form a united image on the retinæ of the eyes, the same effect is produced as if the book itself were looked upon.

This diagram represents the outlines of a near ob-

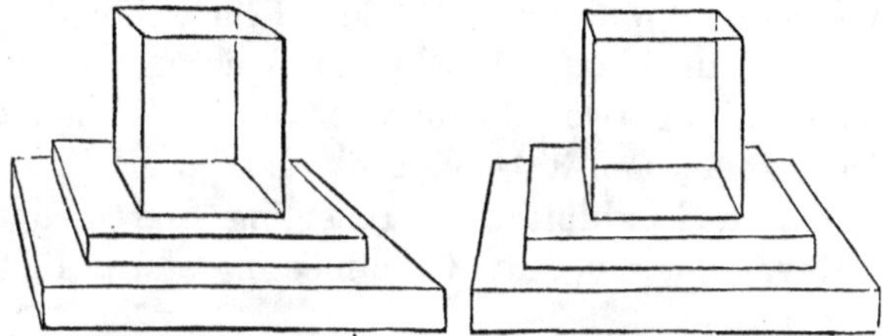

ject, as seen by each eye separately. The one on the right hand shows it as seen with the right eye, and the other as it looks with the left eye; and if both drawings be combined into one image, it stands out in bold relief. This may be done without any instrument, by squinting at them; but the effect is more readily and far more agreeably produced by the Stereoscope, so named from the Greek words *στερος*, solid, and *σκοπεω*, to see.

Professor Wheatstone claims to be the first who contrived an instrument to illustrate this effect of binocular vision, and he also claims to be the first who brought to notice the different appearances of objects seen with each eye separately. Sir David Brewster, however, disputes, on behalf of Mr. Elliot, of Edinburgh, Professor Wheatstone's claim to the invention of the first stereoscopic instrument; and he has shown that the difference of vision with each eye was remarked by Galen, 1,700 years ago; that it was noticed by Leonardo da Vinci in 1500, and formed the subject of a treatise by a Jesuit, named Francis Aquilonius, in 1613; and that it was a well-known phenomenon of vision long before it was mentioned by Professor Wheatstone.* Mr. Elliot, though he conceived the idea, in 1834, of constructing an instrument for uniting two dissimilar pictures, did not carry it into effect until 1839, the year after Mr. Wheatstone had exhibited his reflecting Stereoscope to the Royal Society, and at the meeting of the British Association.

Mr. Elliot's contrivance, to which Sir David Brewster is inclined to give precedence in point of date, was very inferior in its effects to the reflecting Stereoscope. It was without lenses or mirrors, and consisted of a wooden box 18 inches long, 7 inches broad, and 4½ deep, and at the end of it was placed the dissimilar pictures, as seen by each eye, that were to be united into one. The view he drew for the purpose comprised the moon, a cross, and the stump of a tree, at different distances; and when looked at in the box, the cross and the stump of the tree appeared to stand out in relief.

The accompanying woodcut represents the original stereoscopic pictures, copied from Sir David Brewster's book; and by looking towards the picture on the left

* "The Stereoscope: its History, Theory, and Construction," by Sir David Brewster.

with the right eye, and on the right-hand picture with the left eye, the two will be seen united, and the cross and the stump of the tree will appear to stand out solidly.

The arrangement of the apparatus, as described by Professor Wheatstone, in his paper read before the Royal Society, consists of two plane mirrors, about 4 inches square, placed at right angles; and the drawings, made on separate pieces of paper, were reflected to the eyes looking into the mirrors at their junction. The diagram is a sketch of this arrangement. In the

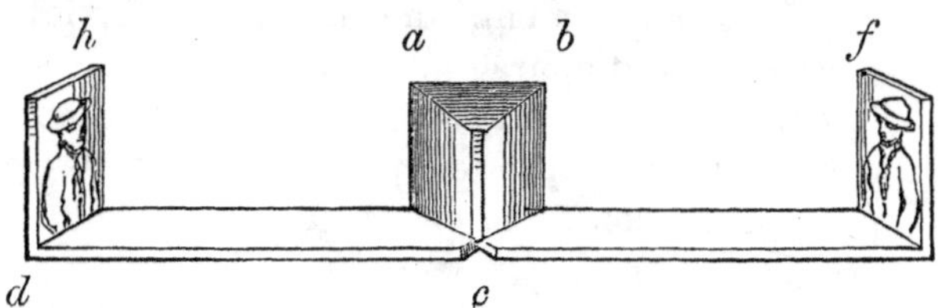

middle of a narrow slip of wood, *d e*, about 12 inches long, the two mirrors, *a b*, are fixed, inclined at the required angle from their line of junction at *c*. Upright pieces of wood, *d h*, *e f*, at each end, are furnished with slides or clips to hold the drawings, which are reflected from the inclined mirrors, and seen in them by each eye separately. Thus, the left eye sees only the picture fixed on *d h*, and the right eye sees the one placed at *e f;* and the two images, being combined at the seat of vision, produce the same impression as a solid body.

It is almost unnecessary to describe the external appearance of the lenticular instrument invented by Sir David Brewster, and explained by him at the meeting of the British Association in 1849. In the best kind of instruments the glasses, through which the pictures are seen, are composed of a single large double-convex lens, divided in the middle, the thin edges being set towards each other, about 2½ inches apart. The more improved instruments, indeed, are made from lenses upwards of 3 inches in diameter, which, being cut into two, and the thin parts being ground flat, are set edge to edge, and from an aperture sufficiently large for both eyes to look through. By this means the instrument suits all eyes, without requiring adjustment, and the field of view is increased. A diaphragm, or partition, placed at the junction of the two lenses, confines the vision of each eye to its appropriated picture, and thus tends to prevent the confusion of images that might otherwise arise.

The object of using semi-lenses is to facilitate the union of the two pictures into one, by looking through the lens towards its edge, instead of through the centre, the image being thus refracted to a different position. This may be easily exemplified by looking at an object steadily through different parts of the same lens. After looking at it with the right eye through the centre, and whilst keeping the axis of the eye in the same direction, move the lens slowly towards the right, so as to bring the edge of the lens opposite the pupil. This movement of the lens towards the right hand will be accompanied by an apparent movement of the image towards the left, so as to bring it to a point between the two eyes. If the experiment be repeated with the left eye, the image will be removed towards the right hand; and thus, by looking at the two stereoscopic pictures through the thin parts of two lenses, the images are superposed and form a single one.

Sir David Brewster attached much importance to the semi-lenses, which have the effect of prisms in refracting the rays of light; but that form of lens is not essential to give apparent solidity to the images; and many of the commoner kind of instruments are now made with ordinary double-convex lenses, and without any partition. With the semi-lens, however, there is less difficulty in uniting the two pictures into one than when an ordinary lens is employed.

In taking photographic pictures for the Stereoscope with a single camera, it is necessary to alter the angle of the instrument after having taken one picture, to direct it to the same object in the angle of vision as seen by the other eye. This method of producing stereoscopic pictures with the same camera is very objectionable when any moving objects are in the field; for they will be in a different position in each, and sometimes disappear altogether from the second picture. The plan adopted by the best photographers is to have two cameras set at the requisite angle to each other, so that both pictures or portraits may be taken at the same time.

At the meeting of the British Association in 1853, M. Claudet endeavoured to establish some rules for the angle at which photographic pictures must be taken, in order to produce the best effect of relief and distance without exaggeration. He observed, that in looking at a single picture with two eyes, there is less relief and less distance than when looking at it with one eye, because in the latter case we have the same effect we are accustomed to feel when we look at the natural objects with one eye; while, if we look at the single picture with two eyes, we have on the two retinæ the same image with the same perspective, which is not natural, and the eyes have not to make the usual effort for altering their convergence according to the plane on which the object observed is situated. This inaction of the convergence of the eyes diminishes the illusion of the picture, because the same convergence for all the objects represented gives an idea that they are all placed on the same plane. The photographic image being the representation of two different perspectives, we must, when we look at them in the Stereoscope, as when looking at the natural objects themselves, converge, more or less, the axes of the eyes. Therefore we make the same effort, and have the same sensation in regarding the combined photographic pictures, as when we look at the objects represented.

Sir David Brewster has suggested various applications of the Stereoscope; viz., to painting, to sculpture and engineering, to natural history, to education, and to purposes of amusement. The latter is the principal purpose to which the instrument is at present applied; and some of the many ways in which it may contribute to delight the spectator are pointed out in Sir David Brewster's book.

"For the purpose of amusement," he observes, "the photographer might carry us even into the regions of the supernatural. His art enables him to give a spiritual appearance to one or more of his figures, and to exhibit them as 'thin air,' amid the solid realities of the stereoscopic picture. While a party are engaged with their whist or their gossip, a female figure appears in the midst of them with all the attributes of the supernatural. Her form is transparent; every object or person beyond her being seen in shadowy but distinct outline. She may occupy more than one place in the scene, and different portions of the group might be made to gaze upon one or other of the visions before them. In order to produce such a scene, the parties which are to compose the group must have their portraits nearly finished in the binocular camera, in the attitude which they may be supposed to assume if the vision were real. When the party have nearly sat the proper length of time, the female figure, suitably attired, walks quickly to the place assigned to her, and after standing a few seconds in the proper attitude, retires quickly, or takes as quickly a second, or even a third, place in the picture, if it is required, in each of which she remains a few seconds, so that her picture in these different positions may be taken with sufficient distinctness in the negative photograph. If these ope rations have been well performed, all the objects immediately behind the female figure, having been previous to her introduction impressed upon the negative surface, will be seen through her, and she will have the appearance of an aërial personage, unlike the other figures in the picture."

It is in the foregoing manner that the remarkable stereoscopic effect of "Sir David Brewster's ghost" is produced, a representation of which is given in the next page.

Sir David Brewster mentions many other curious applications of the Stereoscope, among which are the dioramic effects of pictures seen alternately by reflected and by transmitted light; a daylight view being apparently lighted up artificially in the night, by seeing it at one time with the light reflected from the surface, and then excluding the light from the front, and viewing it as a transparency.

One of the most interesting effects of the Stereoscope has been recently produced by Mr. De la Rue, who has contrived the means of giving apparent rotundity to the surface of the moon, as viewed through a powerful telescope. The disc of the full moon, however highly magnified, presents, as is well known, the appearance of a flat surface, with the lights and shadows marked seemingly on a plane. Owing to the great distance of that luminary, there is no variation in its appearance, whether it be looked at with one eye or with the other, therefore it seems removed beyond the operation of the ordinary cause

of stereoscopic effects. Nevertheless, Mr. De la Rue has taken photographs of the moon which, when placed in the Stereoscope, combine to form a solid-looking globe, on which all the lights and shadows are distinctly and beautifully delineated. He has produced this effect by taking his photographs at different periods of the year, when there is a slight variation in the direction of the moon's face to the earth; and by combining these separate photographs into one image in the Stereoscope, the form of the moon appears as convex as the surface of an artificial globe.

M. Claudet, who is one of the most successful photographers in the metropolis, has contrived an arrangement which he calls a "Stereomonoscope," by which the appearance of solidity is communicated to a single image formed on a screen of ground glass. The screen of ground glass has a black back, and is placed in the focus of a lens in an ordinary camera obscura, wherein the image may be seen by looking down upon it. The particles of the roughened glass reflect to each eye different parts of the image focused on the screen, and by this means a similar effect is produced as when two dissimilar pictures are looked at through a stereoscope instrument. One great ad vantage of this arrangement is that several persons may look at the image at the same time.

Mr. John Sang, of Kirkaldy, has very recently imparted stereoscopic effect to copies of paintings and engravings, the flat surfaces of which were previously thought to defy any such application of the Stereo scope. The means he employs of doing so are at present kept secret, but he has shown its practicability by copying, on wood engravings, Mr. George Cruikshank's series of "The Bottle." In some respects this process seems almost more wonderful than the original Stereoscope, for it gives solid form and apparent substantiality to the mere creations of the artist's pencil.

THE ELECTRIC TELEGRAPH.

No application of science has so completely realized the visions of fancy as the Electric Telegraph. So closely, indeed, does the real of the present day approach to the ideal of ages past, that it might be supposed the narratives in the tales of faëry land were true records of the inventions of former times, and that the combined efforts of inventive genius during the last half century were but imitations and reproductions of what had been successfully accomplished "once upon a time." There is also an intermediate period—between the indefinite of faëry tales and the positive of scientific history—in which sympathetic tablets and magical loadstones, scarcely less mythical, are stated to have been invented; and the individuals are named who thus paved the way for instantaneous communication between all parts of the world.

The Jesuits of the sixteenth and seventeenth centuries took the place of the magicians of the Middle Ages. In the seclusion of their monasteries, they speculated on the mysterious powers of Nature, then partially revealed to them, and shadowed forth images of their possible applications. It is to a vague speculation of this kind that we may attribute the notice given by Strada, in his "Prolusiones Academicæ," of the sympathetic magnetic needles, by which two friends at a distance were able to communicate; though the then fanciful idea has been literally realized. A still more extraordinary foreshadowing of one of the most recent improvements of the Electric Telegraph was the transference of written letters from one place to another by electric agency. This is said to have been accomplished by Kircher, who, in his "Prolusiones Magneticæ," describes, though very vaguely, the mode of operation. But even admitting that there were substantial foundations for these imaginary phantasms, that would not in the least detract from the merit of those who, following closely the footsteps of scientific discovery, have successfully applied the principles unfolded by the investigations of others, and by their own assiduous researches. Thus, whilst steam navigation was facilitating the means of intercourse over rivers and seas, and whilst railways and locomotive engines served to bring distant cities within a few hours' journey of each other, another source of power, infinitely more rapid in its action than steam, has been made to transmit intelligence from place to place, and from one country to another, with the speed of lightning.

The plan of making communications by signals has been in operation from time immemorial; the beacon lights on hills having served in ancient as well as in modern times to give warning of danger, or to announce tidings of joy. Such simple signals were not capable of much variety of expression; but even beacon lights might be made to indicate different kinds of intelligence, by multiplying the number of the fires, and by altering their relative positions. It was not, however, till the invention of telegraphs that anything approaching to the means of holding regular communication by signals was attained. The semaphore of the brothers Chappe, of France, invented by them in 1794, was the most perfect instrument of the kind, and was generally employed for telegraphic purposes, until it was supplanted by the Electric Telegraph.

The semaphore consisted of an upright post, having arms on each side, that could be readily extended, at any given angle. The extension of these arms on one side or the other, either separately or together, and at different angles, constituted a variety of signals sufficient for the purposes of communication. The semaphores, erected on elevated points, so as to be visible through telescopes, signalled intelligence slowly from one station to another, till it reached its ultimate destination; and thus—daylight and clear weather permitting—brief orders could be sent from the Admiralty to Portsmouth in the course of a few minutes. But the communication was liable to be interrupted by fogs, as well as by nightfall.

A remarkable instance of the imperfection of sight telegraphs occurred during the Peninsular War. A telegraphic despatch, received at the Admiralty from Portsmouth, announced—"Lord Wellington defeated;"—and then the communication was interrupted by a fog. This telegraphic message caused great consternation, and the utmost anxiety was experienced to learn the extent of the supposed disaster. When, however, the fog dispersed, the remainder of the message gave a completely opposite character to the news, which in its completed form ran thus: "Lord Wellington defeated the French," &c.

Some better means of transmitting important intelligence was evidently wanted; for not only was the semaphore liable to frequent interruptions by the weather, but its action was very slow, and the frequent repetitions from station to station increased the risk of blunders.

The instantaneous transmission of an electric shock suggested the means of communicating with greatly increased rapidity; and when it was ascertained, by experiments made by Dr. Watson at Shooter's Hill, in 1747, that the charge of a Leyden jar could be sent through a circuit of four miles, with velocity too great to be appreciable, the practicability of applying electricity for conveying intelligence became at once apparent.

Of the many means by which this object was attempted to be accomplished, it will be only possible, in this general survey, to notice those that mark the first steps of the invention, and the most important of those that have accompanied its progress to the present time.

The first method that suggested itself was to transmit signals by means of pith-ball electrometers. When, for instance, two pith-balls are suspended from a wire that is made to form part of an electric circuit, the electricity communicated to the balls causes them to diverge, and when the electricity in the wire is discharged, they immediately collapse. This action of pith-balls, when electrified, was the simplest mode known of making telegraphic signals, and it was accordingly adopted by several of the early inventors of Electric Telegraphs. The first person who proposed to apply it for that purpose was M. Lesage, of Geneva, in 1774. His plan was to form 24 electric circuits by as many separate wires, insulated from each other in glass tubes; and to place in the circuit, at each communicating station, an equal number of pith-ball electrometers. Each electrometer was to represent a letter of the alphabet, and they were to be brought into action by an excited glass rod. When a communication was to be made, the wires connected with the separate galvanometers were to be charged alternately with electricity by the excited rod of glass; and the person at the receiving station, by noticing which of the electrometers were successively put into action, could spell the words intended to be communicated.

By the means thus proposed, correspondence could have taken place at only short distances, for the charge of an excited glass rod would have been too feeble to produce any sensible effect on the electrometers had the length of the circuit been considerable. This difficulty might have been overcome by substituting the charge of a Leyden jar for the excited glass; but the more serious obstacle to the use of such a telegraph would have been the cost, and the difficulty of insulating the 24 wires required to work it.

Most of the early telegraphic inventors encumbered their inventions with the same obstacle, as they seemed to consider it necessary to have a separate circuit for each letter of the alphabet. It was not so however, with all; for M. Lomond, a Frenchman, who ranks second in the list of telegraphic inventors, modified the principle of M. Lesage, so as to enable him to work with only two wires and one electrometer at each station. With the experience since gained in the application of the needle telegraph, such an arrangement seems very simple, and we are inclined to wonder that it was not generally adopted, especially after M. Lomond had shown the way.

To produce all the requisite signals with a single pith-ball electrometer, it was necessary to vary the durations of each divergence, and to combine several to form a single symbol. Thus, suppose that a single divergence of the pith-balls for a second was understood to signify the letter *A;* one divergence, followed by an immediate collapse, by discharging the electricity, might signify *B;* two prolonged divergences might signify *C*, and two short ones *D;* and by thus increasing the number and varying the divergences of the two pith-balls, all the letters of the alphabet might be indicated.

A still more direct method of representing the letters of the alphabet was proposed by M. Reizen in 1794, by the application of the means frequently adopted for exhibiting the light of the electric spark. The charge of a Leyden jar was sent through strips of tin foil, pasted on to a flat piece of glass, so as to form several lines, joined at the ends alternately into a continuous circuit. Interruptions were made in the foil by cutting small portions away, at which points brilliant sparks appeared when the jar was discharged. As the interruptions were so contrived as to form letters, and the strips of tin foil were all arranged separately on a long pane of glass, any letter required could be distinctly made visible by discharging the jar through that particular circuit. To produce all the letters of the alphabet in this manner, a separate circuit was required for each.

Another plan, far less feasible, and scarcely deserving of notice, excepting for its peculiarity, was proposed in the following year by M. Cavallo, who suggested the setting fire to combustibles, or the explosion of detonating substances, as the means of signalling intelligence. About the same time several attempts were made by electricians in Spain to transmit signals by electricity, but their plans were not more practicable than those already mentioned, and depended for their effects on the discharge of Leyden jars.

The discovery of voltaic electricity at the beginning of the present century was an important step in the progress of the Electric Telegraph, though several years elapsed before the applicability of the discovery for that purpose became known; and it was not fully appreciated till within the last twenty years.

The electricity generated by the voltaic battery is far greater in quantity than the most powerful electrical machine can excite, whilst its intensity is so feeble that it cannot pass in a spark through the smallest interval of air. It presents, therefore, much less difficulty in the insulation of the wires than frictional electricity, whilst the rapidity of its transmission is for practical purposes equally efficient. The electricity generated by the voltaic battery being great in quantity and feeble in intensity, it is capable also of effecting chemical decomposition and of imparting magnetism, both of which properties have proved eminently useful in perfecting the Electric Telegraph.

The first application of voltaic electricity to telegraphic purposes was made by Mr. Soemmering in 1809. The signals of his telegraph consisted of the bubbles of gas arising from the decomposition of water, during the action of the electric current. His apparatus consisted of a small glass trough, filled with acidulated water, through the bottom part of which were introduced several gold wires corresponding to the letters of the alphabet. The instant that an electric current was sent through any two of the wires, by making connection with a voltaic battery at the transmitting instrument, bubbles of hydrogen gas rose from one of the gold wires, and bubbles of oxygen gas from another; and as the volume of hydrogen gas, liberated during the decomposition of water, exceeds by sixteen times that of the oxygen, it was easy to distinguish them. In this manner all the letters of the alphabet could be indicated by using 24 wires. The object of having gold wires in the decomposing trough was to prevent the oxidation of the metal; for had copper, or any other metal that combines with oxygen, been employed, the points of the wires would soon have become corroded.

This telegraph of Soemmering's, though not adapted for practical application in the form he presented it, on account of the number of wires required for the purpose, was nevertheless superior to any that had previously been invented; and by a little modification it might have been made a perfect instrument, capable of transmitting messages by means of only two wires. Such a modification of the instrument was proposed by M. Schweigger, twenty years afterwards; the only thing required being the adoption of a code of symbols, by means of which all the letters might be indicated by combinations of the four primary signals that are obtainable by two wires, as is at present done by the needle telegraph in common use. At that time, however, the discovery of the magnetic properties of the electric current, and other improvements in the means of communicating, superseded for some years the use of signals made by electro-chemical decomposition.

The next important step in the progress of telegraphic invention, after that of Mr. Soemmering, was made by Mr. Ronalds, who in 1816 succeeded in making a perfect apparatus, that transmitted every requisite signal with the use of only a single circuit. In the agent employed, however, there was a retrogression to frictional electricity and the pitch-ball electrometer, for at that time the property which a voltaic current possesses of deflecting a magnetic needle had not been discovered.

Mr. Ronalds's plan was to have, at each communicating station, a good clock with a light paper disc fixed on to the seconds wheel, on which were marked all the letters of the alphabet, and the ten numerals. Only so much of this disc was exposed to view as to show a single letter at a time, through a small aperture, as the seconds wheel revolved. The clocks at the corresponding stations were set exactly together, so that the same letter was exposed to view at each instrument at the same instant. A pith-ball electrometer, connected in a single circuit with the transmitting station, was kept distended during the transmission of a message by charging the wire from an electrical machine; and when the letter required to be indicated appeared at the aperture of both instruments, the operator at the transmitting instrument instantly discharged the electricity of the wire by touching it, and thus caused the pith-balls to collapse. In this manner the person at the receiving station, by attentively watching the pith-balls, and noticing the letter that appeared at the instant of collapse, could read the messages signalled.

Mr. Ronalds so far perfected his invention, that it worked accurately, though slowly, through eight miles of wire insulated in glass tubes. Having thus succeeded in putting into action his single-wire telegraph, Mr. Ronalds sought the patronage of Government for its practical adoption, such a notion as that of establishing a telegraph for commercial purposes not being at that time entertained. For a length of time his application received no attention, and when at length the Lords of the Admiralty condescended to answer, they sent Mr. Ronalds, as the reward for his ingenuity, and as compensation for the time and money bestowed in perfecting the invention, the expression of their opinion—that "telegraphs are of no use in time of peace, and that during war the semaphore answered all required purposes"! This reply, so characteristic of the manner in which Government *employés* generally regard anything new to which their attention is solicited, completely disheartened Mr. Ronalds. He abandoned the Electric Telegraph to its fate; and having gone abroad, he returned some years later to find that, notwithstanding the *dictum* of the Lords of the Admiralty, telegraphs are of great use in time of peace as well as of war, and that the old semaphore had been entirely superseded by the means of transmission he had indicated twenty years before. Mr. Ronalds has since received a small pension, not however as a reward for his ingenious telegraph invention, but for his services in other departments of science.

The discovery of the magnetic property of an electric current by Professor Œrsted, in 1818, was most important in the subsequent progress of telegraphic invention, though it was not applied in a practical manner till nearly twenty years afterwards. In 1820, indeed, M. Ampère submitted to the Academy of Sciences at Paris a telegraphic instrument for the transmission of signals by the deflection of needles, but he adopted the impracticable plan of the earliest inventors, of having a separate wire for each letter or the alphabet. A much more important contribution to telegraphic invention by M. Ampère was the discovery of electro-magnets, which act an important part in many recent electric telegraphs.

As the magnetic properties of a voltaic current are extensively applied in electric telegraphs, it is desirable briefly to explain the nature of the action of voltaic batteries before proceeding farther with the history of the invention.

To excite a current of voltaic electricity, it is usual to employ a series of zinc and copper plates, arranged alternately in separate jars; or, what is now most common, in cells of gutta percha, separated from each other in a gutta percha trough. The cells are nearly filled with diluted sulphuric acid, and a wire is attached to each end of the trough; one being connected with the last zinc plate, and the other with the last copper plate of the opposite ends of the trough. When these wires are brought into contact, electricity is instantly generated by the action of the acid on the zinc plates. The electricity excited by the action on the zinc in one cell is carried on to the next, and that again excites and transfers an additional quantity to the third cell, thus increasing in intensity to the last pair of plates in the series. The *electric current*, as it is called, passes along the wire, and whether the wire be one yard, or whether it be a hundred miles long, the generation of electricity takes place the instant that the circuit is completed, and ends the instant that the circuit is broken. There is this difference, however, in the transmission of electricity through a long and through a short circuit, that in the former case the increased resistance offered by the length of the wire greatly diminishes the quantity of electricity transmitted though it does not perceptibly retard the velocity.

When a balanced magnetic needle is held above a short thick copper wire whilst it is transmitting an electric current, the needle is deflected from its natural position, and inclines either to the right or to the left, according to the direction in which the current passes. If, for instance, the north pole of the needle be pointed towards the copper pole of the battery, it will be deflected towards the east, but if the direction of the battery current be reversed, the deflection will be towards the west. The effect instantly ceases when the current is interrupted by breaking connection with either pole of the battery. The copper wire, though under ordinary circumstances incapable of being rendered magnetic, thus becomes endowed with strong magnetic properties when it is transmitting an electric current, and acts on the magnetic needle in the same manner as if there were an immense number of small magnets placed along the wire across its diameter.

The magnetic property of an electric current, first discovered by Œrsted, was applied by M. Ampère to impart magnetism to iron, by coiling a length of copper wire round a bar of iron, taking care to cover the wire with an insulating substance, so that when an electric current was transmitted the electricity might not pass through the iron. Coils of copper wire, covered with cotton or silk, can thus impart most powerful magnetism to a piece of soft iron; but it loses its magnetic power the instant that the electric current is interrupted.

The effect of a coil of insulated wire in increasing the magnetic power of an electric current, was applied by M. Schweigger in 1832 to increase the sensitiveness of a suspended magnetic needle. By surrounding a compass needle with several convolutions of covered wire, it was found that the deflections of the needle were much greater and more active; and he thus showed the way to the construction of those delicate galvanometers, which indicate by their deflections the slightest disturbance of electrical equilibrium. Schweigger may, therefore, be considered the original inventor of the Needle Telegraph; and as he pointed out a method of impressing symbols on paper mechanically, by means of electro-magnets, he may be considered also as the original inventor of Recording Electric Telegraphs.

The first near approach to the needle telegraph, now used in this country, was made by Baron de Schilling, who, in 1832, constructed at St. Petersburg an electric telegraph consisting of five magnetic needles. This may be considered as the precursor of the five-needle telegraph, first patented by Professor

Wheatstone in 1837. By the separate deflection of those needles to the right hand or to the left, by reversing the connections with the poles of the batteries, ten primary signals could be obtained; and by bringing two into action at the same time, many more signals might be made than were required for indicating the letters of the alphabet, and they could be appropriated to express several words. For the action of this very efficient telegraph only five wires were required, and the signals being all primary ones, the messages might have been transmitted very quickly.* In a subsequent modification of the telegraph, he contrived to make all the signals with one magnetic needle alone, by repeating the deflections to the right and to the left, as done in the needle telegraph now generally used in England.

Another step made by Baron de Schilling was the invention of an alarum to call attention when a message was about to be sent. Some contrivance of this kind was considered essential in the early days of the practical application of the Electric Telegraph, as no one then contemplated that telegraphic communications would be so frequent as to require a person to be always near the instrument, waiting for the receipt of messages.

* Primary signals are those in which the letter indicated is represented by a single deflection of the needles in either direction. A single needle telegraph can have only two primary signals, one to the right and one to the left; all the other letters being indicated by repeated deflections. In several instances four deflections are required to signal a single letter.

Baron de Schilling's alarum was very simple. One of the magnetic needles acted as a detent which held a weight suspended, and when the needle was deflected, the weight fell upon a bell. The alarums subsequently invented were constructed on the same principle, but instead of employing one of the magnetic needles as a detent, an electro-magnet was used for the purpose, and clock mechanism was introduced to sound a bell continuously, as soon as it was set in action by the withdrawal of the detent. At the present time alarums are not used in the regular stations of the electric telegraph companies; the sound of the needles, as they strike against the ivory rests on each side, being sufficient to call the attention of the clerks, who are in constant attendance.

We have hitherto been enabled to trace, step by step, the advances made at intervals—years asunder—in bringing the Electric Telegraph into practical use; but we are now approaching a time when it becomes difficult to enumerate, and impossible to describe within reasonable compass, the numerous inventions that were patented and otherwise made known for giving greater efficiency to that means of communication.

In the early part of the year 1837, the electric telegraphs of Mr. Alexander, of Edinburgh, and of Mr. Davy, were publicly exhibited in London, and excited much attention; though at that time it was not supposed that it would be possible to make use of that means of communication for general purposes. Mr. Alexander's telegraph was the same in principle as those of M. Ampère and of Baron de Schilling' though in some respects not so efficient as either, for its action was slow, and it required a separate wire for each letter of the alphabet. It was considered a great advantage of this telegraph at the time, that it exhibited actual letters of the alphabet, instead of symbols. This was effected by having the twenty-six letters painted on a board, and concealed from view by a number of small paper screens, which were attached to magnetic needles. When any of the needles was deflected by sending an electric current through the surrounding coil, the screen was withdrawn and exposed the letter behind. Twenty-six keys, resembling those of a pianoforte, were ranged in connection, one with each wire, and on pressing down any one of the keys, contact was made between the battery and the wire connected with its associated magnetic needle; and in this manner, messages might easily be transmitted and read. The objections to this telegraph, in the form in which it was exhibited, were not only the impracticability of laying down and insulating so many wires, but the paper screens attached to the needles impeded their action, and rendered the transmission a very slow process. It is questionable, indeed, whether that telegraph could have been worked at all through a circuit of many miles.

Mr. Davy's telegraph was similar to that of Mr. Alexander's, though much more compact and better arranged. The letters were painted on ground glass, lighted behind, so that when the screens were withdrawn the letters were seen in transparency.

Professor Wheatstone, who had for some previous years been endeavouring to perfect a practical electric

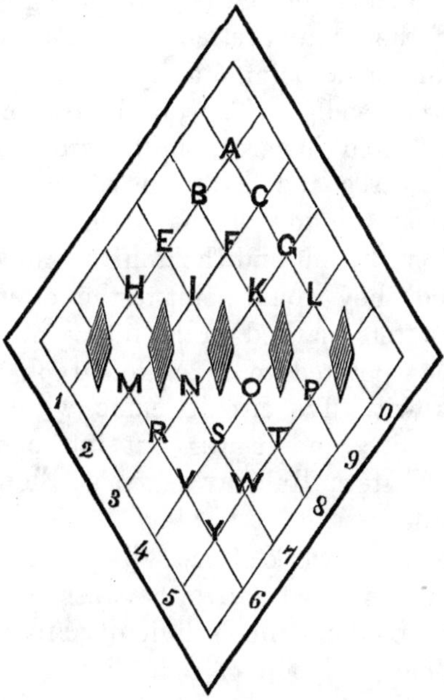

telegraph, took out his first patent in 1837. It closely resembled in general features the telegraph of Baron de Schilling. It consisted of five magnetic needles, ranged side by side on a horizontal line that formed the diameter of a rhomb. The needles were suspended perpendicularly, being kept in that position by having the lower ends made slightly heavier than the upper. The rhomb was divided into thirty-six equal parts by ten cross lines, and the needles were placed at the points where the lines intersected, as shown in the diagram.

At each intersection, and along the boundary lines of the rhomb, letters were marked, any one of which might be pointed at by the combined action of two of the needles. Thus, if the two extreme needles were deflected inwards, one towards the left and the other towards the right, they would point to the letter *A* at the top of the rhomb. If the extreme needle on the left and the fourth one were similarly deflected, they would point to the letter *B;* and thus all the letters marked on the intersections of the lines could be pointed to. A telegraph that could be worked with five circuits came within the range of practicability, and it was put into operation on the Great Western Railway as far as Slough, a distance of 18 miles.

When the work of actually making communication by insulated wires between places far apart came to be done, much difficulty arose as to the best and cheapest mode of doing it. The plan first attempted was to surround the wires with pitch, and to bury them in a trench in the ground. But this was found to be attended with great inconvenience, for the pitch cracked, and electric communication was established between the adjacent wires. The method of suspending the wires on posts was, we understand, suggested by Mr. Brunel, who had seen wires so suspended for other purposes on the Continent, and he recommended it to Mr. Cooke for the Electric Telegraph. The plan was tried with success, and was generally adopted by the Electric Telegraph Company in extending their lines over the country. We shall have occasion to revert to this practical part of the subject, when describing more particularly the means of making communication from one place to another.

In continuing the history of the invention, as regards the different modes by which communications are transmitted along the insulated wires, the next telegraphs that deserve notice are those of Dr. Steinheil, which became known also in 1837. One of his telegraphs made the signals by sounds, produced by magnetic needles striking, when deflected, against bells of different tones. By another telegraph of his invention the symbols were marked upon paper by small tubes holding ink, fixed to the needles. In this manner the letters of the alphabet were indicated by dots upon a strip of paper, kept slowly moving by clock mechanism. This telegraph could be worked by a single circuit; and it appears that Dr. Steinheil was the first who discovered, or at least who practically applied, the conducting power of the earth for the return current. Each circuit, therefore, consisted of only a single wire; the wire that had been previously used to complete the circuit being superseded by burying in the earth, at each terminus, a small copper plate. Dr. Steinheil also introduced the use of galvanized iron wire. An electric telegraph of this construction was put into operation at Munich, through a distance of 12 miles.

In the following year Messrs. Cooke and Wheatstone so far simplified the arrangements of their needle telegraph as to make all the requisite signals with two needles. With a single combined battery and two wires six primary signals are thus obtained; and by repeating the deflections and combining the action of the two needles, all the letters can be readily and quickly indicated. A single needle instrument was invented by Messrs. Cooke and Wheatstone, but as there are only two primary signals, one to the right and one to the left, the deflections are necessarily repeated more frequently, and the transmission is consequently more slow. The accompanying diagram

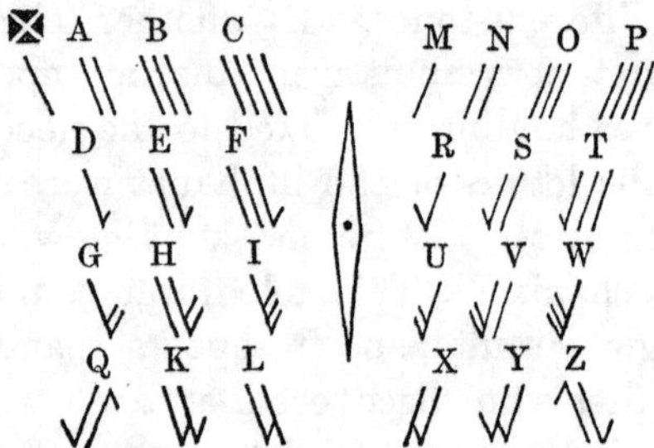

represents the alphabet of the single needle instrument. The deflections for each letter commence in the direction of the short marks, and end with the long ones. Thus, to indicate the letter *R*, the needle is first deflected once to the left and then once to the right; and the letter *D* has the deflections reversed, beginning with one to the right and ending with one to the left. In no instance does it require more than four deflections to indicate a single letter, yet the transmission with the double needle is found so much quicker that the single needle instrument is only rarely used.

At the end of each word, it is customary for the clerk at the receiving station to indicate, by a deflection of the needle to the right, that he understands, or by a deflection to the left, that he does not understand, and in the latter case the word is repeated. In the early days of the Electric Telegraph, the transmission of 40 letters a minute with the double needle instrument was considered quick work; but the practised clerks will now transmit one hundred letters in that time, which is as fast as any person can write with pen and ink.

Since the invention of the double and single needle telegraphs there have been many modifications in the instruments, to make them work more promptly and with less vibration; but in all essential parts the telegraphs of Messrs. Cooke and Wheatstone remain unaltered, and continue to be generally used in this country.

Of the numerous other telegraph instruments that have been invented since 1837, that of Mr. Morse is in most general use, especially on the Continent and in America. Mr. Morse, indeed, claims to be the first inventor of a practical Electric Telegraph; for, according to his statement, he, in 1832, invented a telegraph, which was in principle the same as the one now in use. It was not, however, till September, 1838, that he made his instrument known in Europe, by sending a description of it with a model to the Academy of Sciences at Paris. Mr. Jackson, an American, disputed with Mr. Morse for the honour of the invention, and when the latter asserted that he had described his telegraph in 1832, to some passengers on board a packet-boat, Mr. Jackson affirmed that it was he who described it on that occasion, and that Mr. Morse, being present, got the idea from him. It is painful and difficult to decide when we find two claimants thus directly in opposition to each other, and mutually preferring charges of falsehood and fraud. The only safe guide in such cases is to refer to the earliest published and authentic descriptions of the inventions; and, following that guidance, the invention of what is called Morse's telegraph must be attributed to him whose name it bears; but we must, according to the same rule, date it several years later than 1832.

Mr. Morse's telegraph is a recording instrument, that embosses the symbols upon paper, with a point pressed down upon it by an electro-magnet. The symbols that form the alphabet consist of combinations of short and long strokes, which by their repetitions and variations, are made to stand for different letters. Thus a stroke followed by a dot signifies the letter *A;* a stroke preceded by a dot, the letter *B;* a single dot, the letter *E;* and in this manner the whole alphabet is indicated, the number of repetitions in no case exceeding four for each letter. The letters and words are distinguished from one another by a longer space being left between them than between each mark that forms only a part of a letter or of a word. The annexed diagram represents the symbols for the whole alphabet.

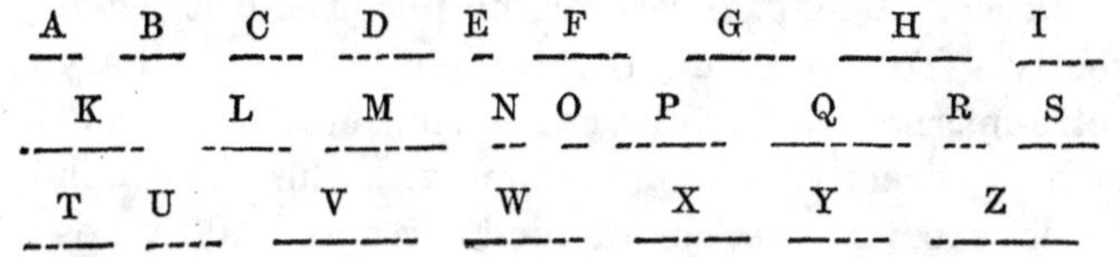

A B C D E F G H I

K L M N O P Q R S

T U V W X Y Z

The mechanism of this telegraph instrument is very simple. The transmitter is merely a spring key, like that of a musical instrument, which, on being pressed down, makes contact with the voltaic battery, and sends an electric current to the receiving station. The operator at the transmitting station, by thus making contact, brings into action an electro-magnet at the station he communicates with, and that pulls down a point fixed to the soft iron lever upon a strip of paper that is kept moving by clockwork slowly under it. The duration of the pressure on the key, whether instantaneous or prolonged for a moment, occasions the difference in the lengths of the lines indented on the paper. A single circuit is sufficient for this telegraph, and a boy who is practised in the use of the instrument will transmit nearly as many words in a minute as can be sent by the double needle telegraph with two wires.

The working of Mr. Morse's telegraph, it will be observed, depends altogether upon bringing into action at the receiving station an electro-magnet of sufficient force to mechanically indent paper. Now the resistance to the passage of electricity along the wires diminishes the quantity transmitted so greatly, that at long distances it would be almost impossible to obtain sufficient power for the purpose, if it acted directly. To overcome that difficulty, an auxiliary electro-magnet is employed. The electro-magnet which is directly in connection with the telegraph wire is a small one, surrounded by about 500 yards of very fine wire, for the purpose of multiplying as much as possible the effect of the feeble current that is transmitted. The soft iron keeper, which is attracted by that magnet, is also very light, so that it may be the more readily attracted. This highly sensitive instrument serves to make and break contact with a local battery, which brings into action a large electro-magnet, and as the local battery and the magnet are close to the place where the work is to be done, any required force may be easily obtained. By this means the marks may be impressed on the paper at distances of 400 miles or more apart.

This is a very efficient and remarkably simple telegraph, and as it operates with a single wire, it has completely supplanted the needle telegraph on the Continent; though the liability to error, common to all manipulated telegraphs, is considerably increased by this mode of transmission, nor can unintelligible signals be indicated and corrected so readily as by the needle instrument.

There have been several modifications of Mr. Morse's telegraph, for the purpose of increasing the rapidity of its action and the distinctness of the marks.

The most important of these was made by Mr. Bain, who in 1847 applied for this purpose the method of impressing the symbols on paper by electro-chemical decomposition. Mr. Davy had, in 1843, taken out a patent for the application of electro-chemical marks to telegraphic purposes, but his method was not sufficiently practical to be brought into use. Mr. Bain adopted an alphabet of short and long strokes, similar to that of Mr. Morse; but instead of making and breaking contact by a key pressed down by the finger, he punched holes in a strip of paper, corresponding in lengths and positions to the marks intended to be transmitted. A small metal spring, connected with the voltaic battery, pressed upon a metal cylinder attached to the telegraph wire, and when the spring and cylinder touched, an electric current was transmitted. The strip of punched paper was placed upon the cylinder so as to interrupt the circuit, excepting in the parts where the apertures allowed the spring to make contact; therefore when the strip of paper was moved along, an electric current was transmitted through the apertures, and it was stopped when the paper intervened. At the receiving station, paper well moistened with a solution of prussiate of potass and nitric acid was placed upon a corresponding cylinder to receive the message, and a piece of steel wire was kept steadily pressed upon it as it moved along. The action of the electric current at the parts where it was transmitted caused the acid to enter into combination with the steel, and the consequent deposition of iron on the paper was instantly converted by the prussiate of potass into Prussian blue. On the parts where the electric current was interrupted no action took place, and thus numbers of short and long marks were made on the paper, corresponding to the lengths of the apertures on the prepared message. A representation of the punched paper for transmitting the word "Bain" is shown in this diagram.

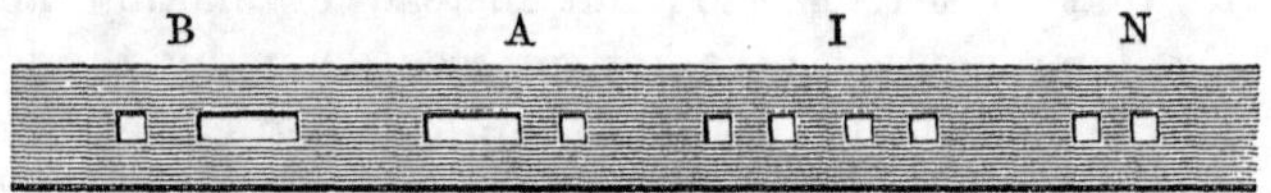

As electro-chemical action takes effect much more rapidly than the mechanical movement of an indenting point, Mr. Bain's telegraph could work much faster than Mr. Morse's. We have been informed that as many as 1,000 letters per minute have occasionally been transmitted by this means from Manchester to London. The disadvantage attending that mode of transmission arises from the tedious process of punching the message preparatory to transmission; and though circumstances may arise in which it would be of great importance to adopt this rapid system of transmission with a single wire, it has been yet but little used in this country by the Electric Telegraph Company, who purchased Mr. Bain's patent for £10,000.

Another modification of Mr. Morse's telegraph, which has been more extensively adopted in England, consists in merely substituting marks made on paper by electro-chemical decomposition for those indented by pressure. It has been found desirable in practice, however, to introduce and auxiliary electro-magnet, called a "picker," for making and breaking contact, by which arrangement the dotted marks can be made by a local battery, and any required amount of electric power be obtained. The marks produced in this manner are more distinct, and are more quickly made, than by mechanical pressure. By a more recent application of Mr. Morse's system, the marks are made on paper with ink flowing through a glass pen, in the same manner as in the telegraph of M. Schweigger, already noticed. As the strip of paper is moved along, a continuous line is thus drawn on the paper. When no signals are being transmitted the line is straight, but when an electric current is sent through the wire, it brings into action an electro-magnet, which attracts the penholder on one side, and alters the direction of the mark. The transmission is effected by making and breaking contact with a key, and the continuance of the divergence of the mark from its normal direction is regulated by the duration of pressure on the key. The symbols are thus made by deviations from the straight line, of different lengths and of varied combinations. Practical application alone can determine whether this mode of making the marks possesses any advantage over Mr. Morse's original plan. The patent for this telegraph was granted to Mr. Wilkins in 1854, but a similar instrument, applied to the notation of astronomical observations, was shown in the American department of the Great Exhibition of 1851.

The recording telegraph instruments hitherto noticed impress on the paper only hieroglyphical symbols, which require long practice to decipher readily. It has, from the first practical application of the invention, been considered highly desirable that the letters of the alphabet should be indicated and printed in their proper forms, so that the momentary transmission of an electric current should leave behind a durable impression that could be read without difficulty. Professor Wheatstone and Mr. Bain separately attempted to accomplish this desired object by the invention of Printing Telegraphs, which print messages from types. It is a question in dispute which of them was the first to design a telegraph of this kind. In 1845, Mr. Bain had a printing telegraph in operation experimentally on the South-Western Railway, for a distance of seven miles, and we are not aware that Professor Wheatstone ever succeeded in working his printing instruments when separated at a distance from each other. In principle, both inventions were similar. A wheel, into the periphery of which were inserted types of the twenty-six letters, was made to rotate in close proximity to a piece of paper, over which was placed a blackened surface that would leave a mark on the paper when pressed upon. When the required letter came opposite the paper, the type-wheel was stopped and forced against it, so that the letter was impressed, and the black from the interposed surface marked the form of the type. The paper was then moved forward to leave space for the next letter, and thus a continuous message could be printed. The objection to these instruments was the uncertainty of stopping the type-wheel at the proper point, so as to avoid printing wrong letters; and when the instruments became thus irregular, they continued so till they were again adjusted. This difficulty has since been overcome; and by the combined efforts of Mr. House in America, and of Messrs. Brett in this country, the printing telegraph has attained a high degree of perfection. The mechanical arrangements of the instrument, though very complex, consist essentially, like those of Mr. Bain and Professor Wheatstone, in having a type-wheel, which, by the action of the operator at the transmitting instrument in making and breaking contact, moves or stops at the required point, and the letters are printed by forcing the paper against the type by an electro-magnet. The movements of the type-wheel are regulated by an electro-magnet, and one great improvement introduced by Mr. Brett prevents the continuance of error, should any be made during transmission, by bringing the type-wheel to its first position after printing each letter, so that if a wrong letter be printed, the subsequent letters will not continue erroneous. This printing telegraph works with a single wire, but its operation is rather slow.

The last recording telegraph we shall notice is the one invented by the author, which transmits copies of the handwriting of correspondents. The communication to be transmitted is written upon tin foil, thinly coated with varnish, with a pen dipped in an ink composed of caustic soda and colouring matter. The alkali detaches the varnish, and when the surface is washed over with a wet sponge, the metal is exposed on those parts written upon, the writing appearing metallic on a dark ground. The message is then placed round a metal cylinder that is connected with the line wire from the receiving station. A brass point, in connection with the voltaic battery, lightly presses on the message as the cylinder rotates, so that the electric circuit is made and broken through the message as it passes under the connecting point, the coating of varnish on the foil being sufficient to interrupt the electric current in those parts where the point is resting upon it. On a corresponding cylinder in the electric circuit, at the receiving station, paper moistened with a solution of prussiate of potass and nitrate of soda is placed to receive the message; and it is pressed upon by the point of a steel wire, in connection with the communicating wire. The accompanying diagram will assist in explaining the arrangement.

The cylinder of the instrument is shown at *a*; *b* is the metal style connected by the wire *g* with one of the poles of the voltaic battery; *o* is the arm which holds the style and serves to insulate it from the rest of the apparatus; *c* is a fine screw on which that arm traverses as the cylinder revolves; *d d* are cog-wheels to turn the screw. The speed of the instrument is regulated by the fan *e*; *f* is the impelling weight, and *h* the wire connected with the distant instrument. The receiving and the transmitting instruments are alike, the only difference between them being that the style of the copying instrument is steel instead of brass wire.

As the cylinder *a* is connected by the wire *h* with the distant instrument, and through it with one of the poles of the voltaic battery, the electric circuit is completed by passing from *g* through the tin foil

message, or through the paper placed on the cylinder. This will be the case whenever the style of the transmitting instrument is pressing on the metallic writing; and at those times the electro-chemical action of the voltaic current will produce a blue mark on the paper of the receiving instrument, by the deposition of iron and its combination with the prussiate of potass. The circuit will in like manner be interrupted whenever the point *b* presses on those parts of the message where the varnish is not removed; and thus, as the two cylinders revolve, there will be a succession of small blue marks on the parts where the writing allows the electric current to pass. As the arms that carry the points traverse on screws, they are drawn along as the cylinders rotate, so as to press on fresh parts of the message and of the paper at each revolution. The steel point would therefore draw a series of spiral lines on the paper, if the electric current were not interrupted; but the interposition of the varnish breaks those lines, and as the point passes over different portions of the letters at each revolution of the cylinder, the marks and the interruptions on the paper correspond exactly with the forms of the letters, and thus produce a copy of the writing placed upon the receiving cylinder, in blue characters on a yellowish ground. Or the message may be written on unprepared tin foil with a pen dipped in varnish; in which case the writing will be copied in white characters on a ground of dark lines, as in the accompanying specimen, *A* being the writing on tin foil, and *B* the message received.

It is essential to the perfect working of the copying telegraph that the corresponding instruments should rotate exactly together. This is effected by an electro-magnetic regulator, which being put in action by one instrument, governs the movements of the distant instrument with the greatest exactness, as proved at a distance of 300 miles.

It might be supposed, as the points must traverse

several times over the same line of writing to copy it, that the process is a slow one; but in consequence of the rapidity with which the cylinders revolve, this is not the case. The ordinary speed is one rotation

in two seconds, and at that rate three lines of writing, containing sixty words, would be copied in one minute, which is three times as fast as an expeditious penman can write.

The advantages proposed to be gained by the copying telegraph, in addition to its increased rapidity of transmission, are the authentication of telegraphic correspondence by the signatures of the writers, freedom from the errors of transmission, and the maintenance of secrecy. As a special means of obtaining secrecy, the messages may be received on paper moistened with a solution of nitrate of soda alone, in which case they would be invisible until brushed over with a solution of prussiate of potass, to be applied by the person to whom the communication is addressed.

Professor Wheatstone has recently contrived an improvement in his index telegraph, which was described by Professor Faraday in a lecture at the

Royal Institution in June last. Its chief merit, however, consists in the beauty of the mechanism, for it is essentially the same as the index telegraphs he and others have previously invented, with the substitution of magneto-electricity for the moving force.

Having now traced the history of the invention of the instruments by means of which messages may be transmitted, it becomes necessary to describe the methods employed for making the electrical connection from one place to another. This part of the electric telegraph system is, after all, the most essential to its efficient working, and bears the same relation to the transmitting instruments that the structure of a railroad does to locomotive engines in the system of railway conveyance.

The fact that an electric current might be sent through a long circuit had been established by Dr. Watson, in conjunction with other Fellows of the Royal Society, in 1747, when they sent the charge of a Leyden jar through two miles of wire, supported upon short sticks driven into the ground; the wire at each terminus being connected with the earth for the return current. This method of insolation and conduction fully answered the purpose, and served to determine the great velocity with which electricity is transmitted, for no perceptible interval occurred between the discharge of the Leyden jar at one end of the circuit, and its effect at the other extremity.

Mr. Ronalds made the next experiment on an extensive scale, by insulating eight miles of wire in glass tubes, the wire being carried backwards and forwards for that distance on his lawn at Hammersmith. That mode of insulation was found very efficient. It was, indeed, too perfect, for the difficulty arose of discharging the electricity from the wire after the charge had passed through it.

The length of telegraphic communication established at Munich, in 1837, by Dr. Steinheil, was an important practical advance in the system of extending and insulating the wires, and deserves consideration, not only from the extent to which it was carried into practical operation, but from the circumstance that the earth was employed to form the return circuit. The wires appear to have been carried through the city by extending them from the church towers and other elevated buildings. That plan, indeed, presents so many facilities for passing telegraph wires through towns, that it is not improbable it may be ultimately adopted in this country.

Though the conducting power of the earth was thus early made use of for one-half of the circuit, the fact seems to have been unknown in England at the time of laying down the telegraph wires to Slough in 1845, for a separate wire was then used for the return current. Some years afterwards, indeed, Mr. Bain laid claim to the discovery; but the fact that the conducting power of the earth had been previously applied to the purpose by Dr. Steinheil has been incontestably proved.

In the early stages of the practical application of electric telegraphs in this country, Mr. Cook took an active part in overcoming the numerous difficulties attending the proper protection and insulation of the wires. In the first instance, the plan of burying the wires in trenches was tried, but with very indifferent success, as the asphaltum and other resinous substances with which it was attempted to insulate them were inadequate for the purpose, and allowed the electricity to escape from wire to wire. The method of supporting the wires on tall posts was then adopted by Mr. Cooke, the wires being insulated from the posts at the points of suspension, by passing them through quills. Various improvements have since been made in the insulators, and the plan most in favour at present is to pass the wires through globular earthenware or glass insulators, attached to the posts, as shown in the annexed diagram. The wires themselves are about one-sixth of an inch in diameter; they are made of iron coated with zinc, or galvanized, as it is termed, to protect them from rust.

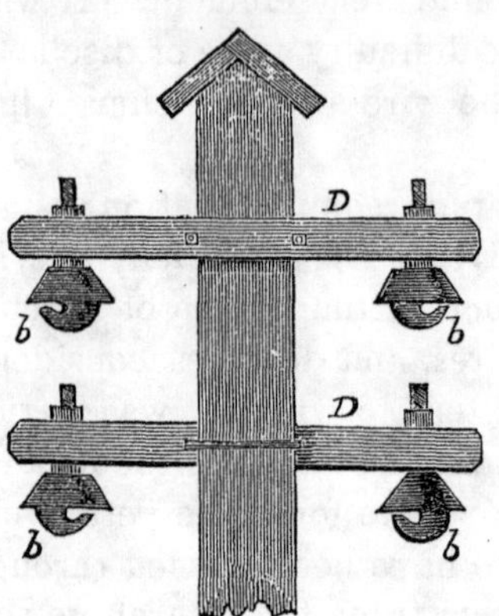

Notwithstanding the great care taken to insulate the wires at the posts, a large quantity of the electricity escapes in wet weather, and returns to the battery without having reached the most distant stations, and thus not unfrequently the communications are interrupted. The author is of opinion that the loss of electricity in wet weather is occasioned rather by communication from one wire to another through the moist atmosphere, than by defective insulation at the posts. In confirmation of this opinion it may be stated, that he has experimentally determined that a working electric current might be transmitted from London to Liverpool, if all the points of attachment were connected by water with the surface of the ground, provided that the rest of the wire were insulated.*

The use of gutta percha as an insulating covering for wire has given rise to a new era in telegraphic communication. Gutta percha is an excellent insulator, and wire covered with two coatings of that material, about one-sixteenth of an inch each, is so far protected, that 100 miles of it immersed in water transmits an electric current from a powerful voltaic battery with very trifling loss. This perfection in insulation has greatly facilitated the establishment of telegraphic communication between England and the Continent. The first attempt to establish a submarine circuit between Dover and Calais took place on the 28th of August, 1850. A single copper wire, about the thickness of a common bell wire, coated thickly with gutta percha, was laid across the English Channel experimentally, without any protection. It proved sufficient for the transmission of an electric current, and several messages were sent through it between Dover and Calais; but it was far too feeble to resist

* "Manual of Electricity," p. 251; and Reports of the Proceedings of the British Association for 1851 and 1854.

the action of the waves, and the following day it was cut through by friction against the rocks, and the communication was stopped.

The plan afterwards adopted for a permanent submarine line was to enclose five similar wires in a hollow iron wire cable. The wires were first slightly twisted, to prevent them from being broken when stretched. They were then covered with hempen yarn, to protect the gutta percha from attrition, and they were thus introduced into the hollow cable, of which they formed the core. The accompanying woodcut represents this structure of the cable; the

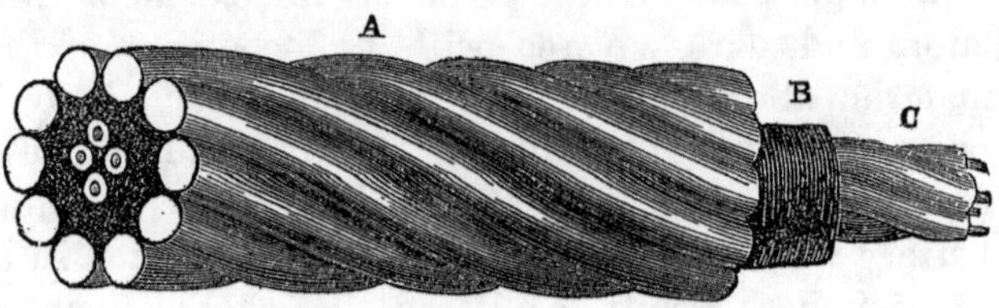

five twisted wires are shown at *C; B* represents the same covered with hemp yarn: and *A* a portion of the completed cable, constructed of thick iron wire galvanized. This cable has now been laid down for seven years, and with perfect success. Its strength has often been severely tested, as it has been sometime drawn up by ships' anchors, and considerably strained; but it has not been broken, and the insulation is almost perfect. The success of this submarine cable has induced the extension of that means of communicating with the Continent, and similar submarine telegraph cables have been laid down from Dover to Ostend, from Harwich to the Hague, from Scotland to Ireland, and across the Mediterranean Sea as far as Malta. The weight and the cost of those cables present a serious obstacle to their adoption in forming a telegraphic communication with America; and when it was determined to attempt to establish electrical connection with the New World, a different form of cable was adopted. The conductor of the

electric current in the Atlantic cable is composed of seven strands of fine copper wire twisted together, the aggregate thickness of which is not greater than the single copper wire of other submarine cables. This fine copper cord is covered carefully with gutta percha; it is then coated with tarred hemp, and is protected externally by an iron wire rope, composed of numerous strands of fine wire. The form and exact size of the cable are shown in the accompanying drawing and section. The central dots in the section are the conducting wires round which are the gutta percha and hemp, and the outer rim represents the iron wire casing.

The successful laying down of so frail a cable, after many failures, affords good ground for hoping that, with the experience already gained, subsequent efforts will prove more satisfactory and much less expensive than this first attempt to establish telegraphic communication with America. The most questionable part of the problem has, indeed, been already solved; for the transmission of electric signals, through that length of submerged wire, was at one time doubted; and though the communication through the present cable has ceased, it has sufficiently established the fact, that telegraphic communication with America is a practicable undertaking.

The excellent insulation obtained by means of gutta percha covered wires has caused a return to the original plan of burying the wires in trenches in the ground. The British and Submarine Telegraph Company make all their communications by that means; the number of coated wires required being enclosed in iron tubes, and laid in the ground along the common roads. That plan is, however, attended with considerable disadvantages. In the first place, the cost of the coated copper wire is more than quadruple that of galvanized iron wire; and though copper, compared with iron, offers only one-seventh part the resistance to the transmission of electricity, yet the thin wire employed is scarcely equal in conducting power to the galvanized iron wire usually supported on posts. The quantity of electricity transmitted is therefore less, and the comparative intensity of it is greater.

Another difficulty attending the use of insulated wires buried in the ground arises from a very peculiar condition of electrical conduction, that could scarcely have been anticipated. The wire, coated with gutta percha, and surrounded externally with water or with moist earth, becomes an elongated Leyden jar; the gutta percha representing the glass, the wire the inside coating, and the water the conducting surface outside. Thus, when electricity is transmitted through such a medium, a portion of the charge is retained after connection with the battery has been broken. This effect increases with the length of the wire and the intensity of the current; and it materially interferes with the working of many telegraph instruments. In some experiments with the copying telegraph at the Gutta Percha Works in the City Road, it was found that through a circuit of 50 miles of wire immersed in water, the mark made by electro-chemical decomposition on paper had a tendency to become continuous; so that instead of ceasing to mark, when the varnish interrupted the current, a line was drawn continuously on the paper, though the stronger marks where the current passed were sufficient to make the writing legible. The retention of the charge was also shown still more remarkably by the explosion of gunpowder by the electricity retained in the wire half a minute after connection with the battery had been broken. It is owing to the retention of the electricity by the wire that the slowness with which the messages through the Atlantic cable were transmitted is to be attributed, and not to the length of the cable. The rate of one word a minute was the average speed of transmission when the first messages were sent through the wire. The effect of the *retardation* of the electric current is comparatively insignificant and were it not for the peculiar action of the surrounding water, the messages might have been transmitted twelve times faster than they were.

The cost of constructing a telegraphic line has greatly diminished with the increased facilities of insulating the wires, and since the expiration of patents, which conferred a monopoly on certain plans of doing so. The cost to the Great Western Railway Company for a line of six wires to Slough, was £150 per mile, with comparatively low and slender posts and very imperfect insulation. The cost of the same number of wires at the present day would not be one-half that sum, with thicker wires and better insulation.

It is customary in England to restrict the suspension of telegraphic wires to railways, from the notion that the protection of railways is necessary to prevent wilful damage to the wires; and as the Electric Telegraph Company have made exclusive arrangements with all the railway companies out of London, the competing telegraph companies have preferred to lay their wires underground rather than incur the supposed risk of damage to the wires if suspended from posts on common roads, though by this means the cost of construction is at least quadrupled. The protection which railways afford is, however, more imaginary than real, for any one inclined to interrupt the communication could easily do so; and if on common roads proper precautions were taken in fixing the posts, and a heavy penalty were imposed on wilful offenders, the common roads and open fields would, there can be little doubt, offer as safe a course for the telegraphic wires as railways.

The conducting power of the earth is now employed by all electric telegraph companies for one-half of every circuit. Thus, whether a communication be sent from London to Liverpool, to Edinburgh, Paris, or Brussels, the moist earth serves to complete one-half of the communication. In the telegraphic circuit between London and Liverpool, for example, the insulated wire is connected at each end with the earth by being soldered to a copper plate, which is buried a few feet underground, so as to insure its being always surrounded with moisture. To improve the connection of this plate with the earth, it is customary to bury with it a quantity of sulphate of copper, the solution of which surrounds the earth-plate with a better conducting liquid than water, and thus extends the connecting surface. The gas pipes or water pipes are sometimes employed for the attachment of the wires instead of an earth-plate, but the latter is generally preferred.

In arranging a telegraphic circuit, the voltaic batteries and the instruments are introduced at breaks in the telegraph wire. The course of the electric current is from the copper end of the battery through the transmitting instrument, then along the wire to the receiving instrument; from that it passes to the earth and is thus returned to the transmitting station, where it completes the circuit by being conducted from the earth-plate to the zinc end of the voltaic battery. The arrangement for completing the circuit will be more clearly understood by reference to the accompanying diagram.

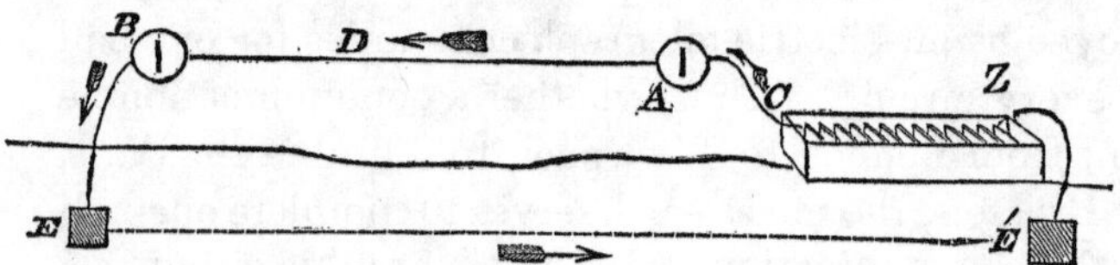

The wire from *C*, which is the copper pole of the voltaic battery, is connected with the instrument *A*; the electric current is then transmitted along the wire *D* to the receiving instrument *B;* thence it is transferred to the earth-plate *E*, passes through the earth to the corresponding plate *E'*, which is connected with *Z*, the zinc pole of the battery. When a communication is returned from *B* to *A*, a similar arrangement is made; the wires connected with the instruments being so arranged as to bring into action a voltaic battery at *B*, and to throw out of circuit the one at *A;* for the connection with the battery is only made when the transmitting instrument is worked.

Since all the electric telegraphs in different parts of the world are connected with the earth, as one portion of the circuit, it might be supposed that the various currents would mingle, and occasion a confusion of messages; but it must be borne in mind that no electric current is formed until a communication be made from one pole of a voltaic battery to the other, and as such communication can only be completed through the insulated wire, the earth-currents cannot mingle, but each one passes to the proper terminus of its respective battery. The accompanying diagram and explanation may serve to remove the difficulty of understanding why the two circuits are maintained quite distinct.

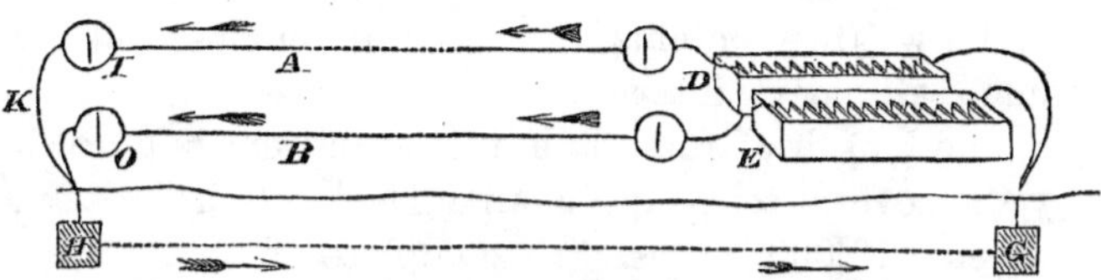

The letters *A B* represent the wires making communications between the batteries *D* and *E*, and the telegraph instruments *I O* at the receiving station. The electricity from the copper end of the battery *D* would be conducted along *A* through the instrument *I*, and by the wire *K* to the earth-plate H. It would be then transmitted through the earth on its return to the battery, in the direction of the arrows, to the other earth-plate *G*, and thence it would find its way to the zinc pole of the battery *D*, and complete the circuit. In the same manner, the electric current from the copper end of the battery *E* would be transmitted through the wire *B*, and would complete its current also by means of the earth-plates *G H*, and would traverse the course indicated by the arrows, and return to the zinc end of *E*. Though both electric currents traverse the same wire from the instruments *I O* to the earth-plate *H*, and are thence transmitted through the earth to a single plate, *G*, at the transmitting station, there is no mingling of currents, the electric current of each battery being kept as distinct as if separate wires were used both for the transmitted and the return current. It would, indeed, be as impossible for the separate currents transmitted from the two batteries to be mingled together, as it would be for the written contents of two letters enclosed in the same mail-bag to intermix.*

The length of telegraph lines at present laid down by the several telegraph companies in Great Britain, exceeds 10,000 miles. To complete those lines required 40,000 miles of wire, and there are 3,000 persons engaged in the transmission of telegraphic intelligence.

In North America there is a direct communication from New York to New Orleans, a distance of 2,000 miles, through the whole length of which wires messages can be transmitted without any break. Wires have also been suspended on lofty posts across the Indian Peninsula, where no railways have been yet laid down. Lines of insulated wire, partly submerged in the sea, partly buried underground, and partly suspended on posts in the air, place London and Vienna in direct communication; and other telegraph lines are in the course of construction, which will unite London with Africa: and a complete network of telegraph wires is spreading over the face of Europe.

It will not be long before this system of communi-

* "Manual of Electricity," second edition, p. 247.

cation is connected with a similar one in America. The failure of the cable already laid down has confirmed the opinion of the author, expressed in papers read at meetings of the British Association for the Advancement of Science, and in his work on Electricity, that the conducting wire should be sufficiently strong to be self-protective, without requiring an external coating of iron wire rope. A conducting copper wire, a quarter of an inch in diameter, covered with gutta percha and tarred hemp, would be more flexible and stronger than the combined cable; and it being a much better conductor of electricity, the rapidity of transmission would be greatly increased.

The effect of the establishment of competing telegraph companies in England has been to diminish the charge for transmitting messages, in some instances to one-fifth of the rate formerly demanded; and when further experience in the construction of telegraphic lines, and the adoption of more rapidly transmitting instruments, have facilitated and improved the means of communication, we may anticipate that correspondence by Electric Telegraph will in a great measure supersede the transmission of letters by post.

ELECTRO-MAGNETIC CLOCKS.

The invention of Electro-Magnetic Clocks closely followed the introduction of the electric telegraph; and Professor Wheatstone, to whom the world is principally indebted, in conjunction with Mr. Cooke, for the perfection and application of the needle telegraphic instrument, claims also to be the original inventor of Electro-Magnetic Clocks. His claim is, however, disputed by Mr. Bain, who asserts that he was the first who conceived the idea of applying the power of electro-magnets to the regulation and movements of clocks, and it must be admitted that he brought the invention into a working state.

In the first stage of the invention, the object attempted to be attained was to regulate several clocks, once an hour—or oftener, if required—so that they might all indicate precisely the same time. For this purpose Mr. Bain took for a standard time-keeper a clock of the best possible construction, placed in circumstances favourable to maintaining accuracy. The minute-hand of his clock, the instant that it pointed to the hour, made connection with a voltaic battery that brought into action a series of electro-magnets attached to the clocks to be regulated; one of them being fixed on the top of each clock. Its momentary action was made to collapse a pair of clippers, which in closing seized the minute-hand of the clock to which it was attached, and brought it to the hour point. Thus all the clocks in the series could be regulated every hour, for the collapse of the clippers pushed the hand forward if it were too late, or thrust it back if it had gained. Mr. Bain contemplated the application of this contrivance to all the public clocks of a town, by having wires laid down in the streets to connect them in one voltaic circuit. Such a plan would, however, have involved greater expense and trouble in its accomplishment than the object seemed to merit; but the regulation of any number of clocks in a large establishment might have been practicable by that means. We are not aware, however, that this mode of regulating clocks by electricity was ever adopted, and it has since been superseded by an arrangement made by Mr. Shepherd, junior, to be presently noticed.

Improving on this first application of electro-magnetism to the regulation of clocks, Mr. Bain afterwards employed the power to keep the clocks in action, so that each clock might be propelled by magnets alone, without any weight, and without the ordinary train of wheels.

Every one acquainted with the mechanism of a clock is aware that the weight communicates motion to a train of wheels, and that the movement is regulated by the vibration of a pendulum, which is acted on by the last wheel of the train. That wheel, called the escapement, is so formed, that each tooth catches in succession into a detent fixed on the pendulum near the point of suspension, which allows one tooth to pass at each double vibration. The pendulum, therefore, governs the movement of the train of wheels by checking the escapement, and allowing the teeth to pass one by one; and as pendulums of given lengths vibrate in given times, if their actions be not interfered with, the clocks will keep regular time. But the pressure of the escape-wheel against the detent, and the consequent friction, prevent the pendulum from acting freely. In the best made clocks there are special contrivances to detach the pendulum as much as possible from the wheels, and likewise to compensate for variations in the leugth of the pendulum by change of temperature.

In the clocks actuated by electro-magnetism, the movement of the pendulum is not maintained by repeated impulses of the escape-wheel, as in ordinary clocks, but by magnetic attraction; an electro-magnet being so arranged as to attract the bob of the pendulum in both directions alternately. In Mr. Bain's arrangement, the bob of the pendulum is formed of a hollow coil of covered copper wire, which, on the transmission of an electric current, becomes magnetic, and it is then attracted by several permanent magnets fixed in a hollow horizontal bar, over which the coil of wire moves. The accompanying diagram will serve to explain more clearly the parts of the clock on which the movement of the pendulum depends.

The pendulum rod, *B*, is made of wood, and the bob, *A*, consists of a hollow coil of thick copper wire covered with cotton, through which the hollow bar, *C C*, passes. Inside that bar there are several permanent magnets, packed on each side of the ends of the coil of wire, the poles of those on one side being the opposite of those on the other. In the diagram only one magnet on each side is represented, *n* and *s*, to prevent confusion. The ends of the coil of wire are attached to the pendulum rod, and they are conducted up it so as to form connection with the wires of the voltaic battery, which are connected with gold studs inserted into a horizontal stage fixed to the clock-case. A small movable bridge, formed of wire, and having the ends tipped with gold or platinum, rests upon the stage, and is shifted from side to side by the pendulum. In these movements the gold points touch and slide over the gold studs in the stage, and thereby make and break contact with the voltaic battery, and alternately send and interrupt an electric current through the coil of wire.

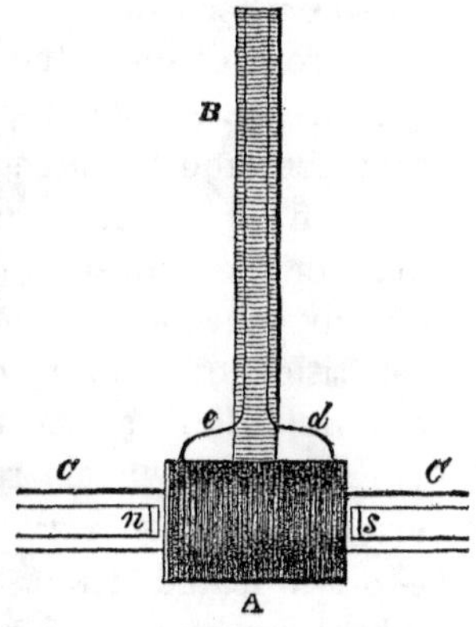

Suppose, for instance, that the pendulum is about to rise to the right towards *s*, at which time the voltaic circuit is completed. The coil is, therefore, magnetic, and is attracted by the permanent magnet in *C.* As the pendulum approaches the end of its swing, it pushes the movable bridge away from the gold studs on which it rests, and thus breaks connection with the voltaic battery, and the pendulum descends unrestrained by the attractive force of the magnets. As the pendulum descends towards its lowest point, it shifts the bridge on to the metal studs on the other side, which are so disposed as to send a current through the coil in a direction opposite to the former, so that the poles of the voltaic battery are reversed, and the attractive force is exerted in drawing the pendulum towards the left hand. In this manner the power imparted to the coil, as the pendulum vibrates to and fro, produces a continuous repetition of the attraction on each side alternately, and maintains a constant action.

The only wheels required in a clock of this kind are those which turn the hands; and the motion is communicated from the pendulum to the seconds wheel by means of a small attached lever, working on a ratchet wheel. The minute and the hour hands derive their movements from the seconds wheel in the usual manner.

The voltaic battery employed to work Mr. Bain's clocks consists of a pair of large copper and zinc plates buried in the moist earth, which excite a sufficient amount of electricity to maintain the motion of the pendulum. A battery of this kind will remain in action a long time, and will serve to keep a clock going for several months. It is, indeed, a near approach to the attainment of perpetual motion, since nothing but the wearing away of the materials, or the accumulation of dust on the connecting points, seems to prevent the realization of that mechanical chimera.

There is a disadvantage attending the arrangement of Mr. Bain's clocks, arising from the attachment of the pendulum to the wheels; and as the moving force is derived directly from voltaic electricity, any variation in the power of the battery causes variation in the lengths of the vibrations, and produces irregularity. For the purpose of remedying these defects, Mr. Shepherd, junior, has adopted an arrangement which detaches the pendulum from the clock movement, and makes its vibrations altogether independent of the varying force of voltaic batteries.

In Mr. Shepherd's arrangement, the impulse of the pendulum is given by successive blows from a spring, which is drawn back and then liberated at each vibration. The hands of the clock are also moved by electro-magnets, by which means the impelling forces and the resistances encountered by the pendulum are always constant. By making the pendulum thus independent of the works, and employing it merely to make and break contract at regular intervals, any number of clocks in the same establishment may be set in motion, and kept exactly together, by a single pendulum.

The large clock over the principal entrance to the Great Exhibition was on this construction. It would have been impossible, with any approach to regularity, to have moved hands of that size, exposed as they were to the wind, unless the pendulum had been independent of such resistances.

Electro-Magnetic Clocks have not yet come into general use, partly owing to imperfections in the battery connections, which occasionally put a stop to their movements, but principally on account of the high prices charged by the patentees. As no trains of wheels are requisite in an Electro-Magnetic Clock, it might be manufactured very cheaply; and when the price is reduced to its proper standard, and the trifling practical defects are remedied, these clocks may possibly supersede others.

ELECTRO-METALLURGY.

The electrotype, electro-gilding and plating, and the other applications of the deposition of metals from their solutions, by the agency of voltaic electricity, had their origin in the chance observation of peculiarities in frequently repeated experiments. In this, as in most other inventions, there are contending claimants for priority; but there is little merit due to any of the first discoverers of the process, who seem to have been guided altogether by accident. It seems strange, now, on observing the extensive use that is made of the deposition of metals, that it should have remained so long unapplied after the principle had been known.

The "revivification," as it was called, of metals from their solutions by voltaic electricity, was known at the beginning of the present century; for, in 1805, Brugnatelli, an Italian chemist, gilded a silver medal by connecting it with the negative pole of a voltaic battery, when immersed in a solution of ammoniuret of gold. It did not occur to him, however, that any use could be made of that mode of gilding, and the experiment had no result.

Nothing further was done, even experimentally, towards advancing the art of electrotyping, until Mr. Spencer, of Liverpool, when experimenting with a Daniell's battery, in 1837, accidentally coated a penny piece with copper; and when the thin sheet of metal was removed, he found on it an exact counterpart of the head and letters of the coin. Even this did not suggest any useful application; nor was it until, by a further accident, a drop of varnish fell on the copper of the negative pole, and showed that no deposition took place on the part so covered, that the idea occurred to him of turning the deposition of the copper to account. The method he adopted of doing so was to cover a copper plate with varnish or wax, and to etch a design through the covering. By then exposing the plate to the action of a solution of sulphate of copper, when in connection with the negative pole of a voltaic battery, the metal was deposited in the lines drawn through the varnish, and a design in relief was fixed to the copper. This slight advance in the art was not made known until it was announced, in 1839, that Professor Jacobi, of St. Petersburg, had made application of the same process. Mr. Spencer, indeed, was forestalled, even in this country, by Mr. Jordan, a printer, who published an account in the *Mechanics' Magazine* for May, 1839, of a method of making copper casts by the deposition of copper from its solution. In the autumn of the same year, however, Mr. Spencer exhibited to the British Association several more perfect specimens of electrotyping, that showed the process might be rendered valuable; and from that time rapid progress was made in bringing it into practical operation in a variety of ways.

The deposition of copper from its solution, when under the action of voltaic electricity, is not produced by the decomposition of the sulphate of copper, as might be supposed, but by the decomposition of the water that acts as the solvent of the metallic salt. Thus, when two platinum wires from the poles of a voltaic battery are introduced into acidulated water, hydrogen gas is disengaged at the wire connected with the negative pole, and oxygen at the other; but when a solution of sulphate of copper is substituted for water, the hydrogen that is disengaged combines with the oxygen that held the copper in solution, and the metal is liberated. The copper thus liberated from its combination with the oxygen is deposited, in a pure metallic state, on the surface connected with the negative pole of the battery.

The simplest illustration of electro-metallic deposition is obtained by immersing a silver spoon and a strip of zinc into a solution of sulphate of copper. So long as the two metals are kept apart, no change takes place on the silver, but on bringing them into contact, voltaic action immediately commences, and a coating of copper is deposited upon the spoon, and adheres firmly to the metal. If the action be continued, and the supply of copper be maintained by the addition of fresh crystals of the sulphate, the coat of copper may be increased in thickness to almost any extent.

The first applications of the discovery were directed

to the copying of medals and coins. An impression of the metal was obtained in fusible metal, which is an alloy composed of tin, lead, and bismuth, melted together in the proportions of two of the latter to one each of the former. This alloy expands on cooling, and thus affords a very sharp impression of the medals; and as it melts at a low temperature, it may be easily removed after the copper coating has been deposited upon it.

An electrotype mould, obtained directly from the medal, is, however, more sharp in its definition than an impression, and is therefore preferable, when circumstances admit of its being so taken. For that purpose, the surface whereon the deposition is to be made is smeared over with sweet oil, or with black lead. It is then carefully wiped with cotton wool, but a minute quantity of the oil will still remain, sufficient to prevent the metal from adhering.

A simple form of apparatus for the electrotype process is shown in the accompanying diagram.

An earthenware jar, *a*, serves to hold the solution of copper, which should be maintained in a saturated state by the addition of crystals of the salt. A porous tube, *b*, holds a rod of amalgamated zinc, to the top of which a binding-screw is soldered. The copper mould or medal, *c*, is suspended in the solution by a wire, which is held tight by the binding-screw, *d*. The porous jar is then filled to the same height as the copper solution in the jar with diluted sulphuric acid, in the proportions of one of acid to twenty of water.

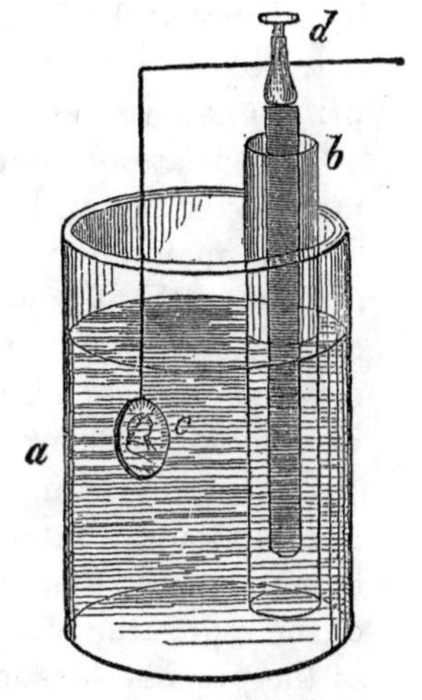

Voltaic action immediately commences, and the copper will continue to be deposited from the solution as long as the supply of fresh crystals of sulphate of copper is continued. In about twenty-four hours the coating of copper will be as thick as a thin card, and it may be then removed. When detached from the medal, it will be found to be an exact counterpart, in the minutest details, of the original; those parts of the medal which are in relief being, of course, the reverse in the mould.

The extreme minuteness and delicacy of the electrotype process is strikingly exemplified in its application to the transference of engraved copper-plates. A highly finished engraved copper-plate has a film of metal deposited over its whole surface, which, when detached, exhibits all the lines that are cut into the copper-plate in relief. That electrotype cast then serves as the mould for further depositions, in which every line in the original engraving is so perfectly developed, that it is impossible to detect a difference in the impressions taken from the two plates. By this means any number of casts may be made and worked from, whilst the original is preserved uninjured. The objection to this application is that the metal deposited is not so hard as the hammered plates, and will not, therefore, bear the wear and tear of copper-plate printing so well as the plates made by hand.

It was at one time supposed that the depositing of metal on surfaces, by voltaic action, might be applied to the manufacture of numerous kinds of copper articles without manual labour. For this purpose, casts were made of plaster of Paris, which were covered with black lead, to give them the property of conducting electricity, and the metal was then deposited upon them. But, independently of the practical difficulties attending the operation, it was found that the metal was not sufficiently hard, and the cost of the requisite voltaic batteries rendered the economy of the process questionable

One of the successful applications of Electro-Metallurgy is founded on the original application of it by Mr. Spencer. As already stated, he covered metal plates with wax, and after scratching through the coating, and exposing the metal, he submitted the plate to voltaic action in a solution of sulphate of copper, and thus obtained a representation, in relief, of the figures cut through the wax; but he does not seem to have thought of the application of this mode of deposition, since adopted, by which engravings in relief are obtained, and printed from with the ordinary letter-press, in the same manner as woodcuts. The name given to this new art is "Glyphography," and it is used with great advantage when the effect of copper-plate engraving is required; for cross lines, which are difficult to cut in wood, can be worked by this method with as great facility as in copper-plate etching.

Another application of Electro-Metallurgy that promises to be useful, is the coating of glass and earthenware vessels with copper, so as to enable them to be placed over the fire without being cracked. A glass sauce-pan might thus be made, which, protected by metal covering, would neither break nor crack when placed upon the fire, because the metal would diffuse the heat over the whole surface, and prevent the unequal expansion of the vessel, which is the cause of the cracking of glass and earthenware when placed upon the fire. A patent was granted last year for a mode of coating earthenware vessels with copper or iron by electro-chemical deposition. The earthenware is first covered either with copper leaf or with bronze powder, to obtain an electrical conducting surface on which the copper can be deposited, and the vessel is then placed in a solution of sulphate of copper, and put in connection with the negative pole of a voltaic battery.

The electrotype is frequently applied with advantage to the preservation and multiplication of objects of art and natural productions, for even the forms of flowers may be in this manner rendered durable; but the most important use that has been made of the process is in plating and gilding. To effect that object, it is necessary to employ a voltaic battery separated from the vessel in which the decomposition takes place. The annexed diagram shows an arrangement of this kind. A single cell of a Daniell's battery, *a*, is connected by wires from its positive and negative poles, with metal rods placed across the decomposition cell, *b*. The articles to be plated are

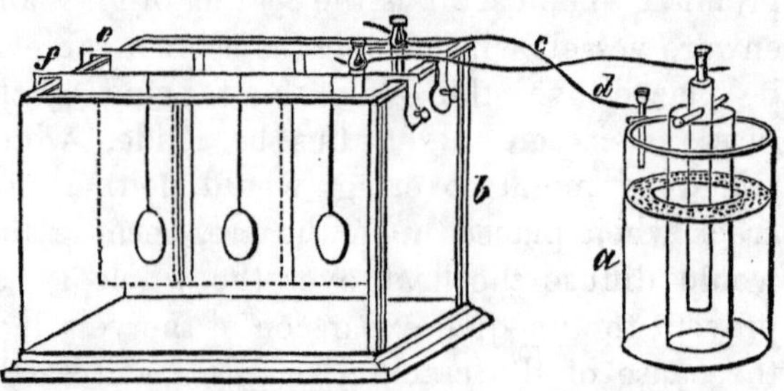

suspended by wires from the metal rod, *f*, and a plate of silver is attached to the rod, *e*. Thus, when the vessel is filled with the silvering liquid, a voltaic current is established, and the deposition is effected on the articles connected with the negative pole.

The menstruum best adapted for electro-plating is a solution of silver in cyanide of potassium. During the process of deposition, the same quantity of metal that is deposited from the liquid on the objects connected with the negative pole of the battery is restored to it, by dissolving an equal quantity from the silver plate connected with the positive pole, and in this manner the solution is maintained at the same strength. Any thickness of silver may be deposited by continuing the process, but about one ounce and a half to a square foot of surface is considered a full quantity. Those parts on which no silver is required to be deposited are covered with varnish or wax, which protects them from the voltaic action.

Where the operation of electro-plating is carried on extensively, decomposing troughs, holding nearly 300 gallons, are employed, and the silver plates in a single trough expose a surface of thirty square feet to the dissolving action of the menstruum under the influence of the voltaic battery.

By the aid of electro-plating the most elaborate designs of the artist in metal can be covered with silver on every part; and a group, finely engraved in copper, may be made to resemble in every particular a work cut out of solid silver. The metal is usually deposited in a granulated state, resembling "frosted" silver, and the parts required to be bright are subsequently burnished; but by the addition of a few drops of the sulphuret of carbon to the solution, the silver may be precipitated perfectly bright.

GAS LIGHTING.

The invention of Gas Lighting had its origin in the earliest times of history; not, indeed, as we now see it, burning at the end of a pipe supplied with gas made artificially, and stored in gas-holders, but blazing from fissures in the ground, supplied from natural sources in the bowels of the earth. The Greek fire-altars are supposed to have been supplied with gas, either issuing from bituminous beds, or made artificially by the priests, by pouring oil on heated stones placed in cavities beneath. Fountains of naphtha, and fires issuing from the earth at Ecbatana, in Media, are mentioned by Plutarch in his life of Alexander, and many other ancient historians record the knowledge of similar instances of natural gas lighting.

In later times, the inflammable gas issuing into the galleries of coal mines, and either exploding when mixed with atmospheric air, or blazing as it issued from fissures in the coal, afforded instances of the natural production of gas, the ignition of which too frequently proved fatal to those in the mines.

A remarkable instance of the issue of inflammable gas into the shaft of a coal mine at Whitehaven, which produced a blaze about 3 feet diameter and

6 feet long, is noticed in the "Philosophical Transactions" of 1733. The part whence the gas issued was vaulted off, and a tube was inserted into the cavity and carried to the top of the pit, where it escaped in undiminished quantity for years, and lighted the country for a distance of several miles. Many experiments were made with this large issue of gas, and it was proposed to conduct it in pipes to Whitehaven, to light the streets of that town, but the proposition was rejected by the local authorities.

In China, naturally produced gas is used on a large scale, and was so long before the knowledge of its application was acquired by Europeans. Beds of coal, lying at a great depth, are frequently pierced by the borers for salt water, and from these wells the inflammable gas springs up. It sometimes appears as a jet of fire from 20 to 30 feet high; and, in the neighbourhood of Thsee-Lieon-Teing, the salt works were formerly heated and lighted by means of these fountains of fire. Bamboo pipes carry the gas from the spring to the places where it is intended to be consumed. These canes are terminated by tubes of pipe-clay, to prevent their being burnt, and other bamboo canes conduct the gas intended for lighting the streets, and into large apartments and kitchens. Thus Nature presents in these positions a complete establishment of gas-light works. *

Though the production of illuminating gas from natural sources had been thus long known, the idea of

* "Treatise on Coal Gas," by Samuel Clegg, jun.

distilling it artificially from coal, for the purpose of illumination, did not occur until the end of the last century. Dr. Clayton, indeed, nearly arrived at the practical application of carburretted hydrogen, in 1737, for he instituted experiments to prove that coal contains gas, nearly similar to that of the "fire damp" in coal mines, and that it burns with a bright flame. At that time, however, the nature of gases was so imperfectly known, that he was unable to do more than discover that coal possesses the property of giving out, when heated, gas that will burn with a bright light.

In the "Philosophical Transactions" of 1739, Dr. Clayton thus describes the effect of the "spirit of coal," obtained by destructive distillation in an iron retort. "I kept this spirit," he says, "in bladders for a considerable time, and endeavoured several ways to condense it, but in vain; and when I had a mind to divert strangers or friends, I have frequently taken one of these bladders, and pierced a hole in it with a pin, and, compressing gently the bladder near the flame of a candle till it once took fire, it would then continue flaming till all the spirit was compressed out of the bladder; which was the more surprising, because no one could discern any difference in the appearance between these bladders and those which were filled with common air."

The first known application of coal gas to illumination was made, in 1792, by Mr. William Murdoch, engineer at the Soho manufactory, to whose great ingenuity the world is also indebted for the invention of the first plan of a steam locomotive engine. * He was at that time occupied in superintending the fitting up of steam engines for the Cornish mines, and his attention having been directed to the properties of gas issuing from coal, he instituted a series of experiments on carburretted hydrogen, the practical result of which was the lighting of his house and offices, at Redruth, with coal gas. The mines at which Mr. Murdoch worked being some miles distant from his house, he was in the constant practice of filling a bladder with coal gas, in the neck of which he fixed a metallic tube with a small orifice, through which the gas issued. The lighted gas issuing through the tube served as a lantern to light his way; and as he thus proceeded along the road with the light issuing from the bladder, the country people looked upon him as a wizard.

The gas was generated by Mr. Murdoch in an iron retort, and collected in a common gasometer, from which it was conducted in pipes to the rooms to be lighted. He also, in an early stage of the invention, contrived a means for making the gas portable, by compressing it into strong vessels; a plan which, a few years since, was adopted by the Portable Gas Company in London.

Mr. Murdoch having proved the practicability of illumination by gas generated from coal, he continued his experiments to facilitate the manufacture of the gas on a large scale, and for its more perfect purifica-

* See article, "Steam Carriages," page 35.

tion. The first public display of its illuminating power was made at the rejoicings for the peace of Amiens, in 1802, on which occasion part of the workshops of Messrs. Boulton and Watt, at Soho, was brilliantly illuminated with coal gas by Mr. Murdoch. In 1805, Messrs. Phillips and Lee, of Manchester, had their extensive cotton mill fitted up with gas apparatus, under the superintendence of Mr. Murdoch, and the quantity of light given out by the burners in all parts of the cotton mill was equal to that of 3,000 candles.*

Notwithstanding these eminently successful trials of gas lighting, the prejudice against innovation prevented, for several years, the extensive adoption of the plan. As every establishment using gas had to make it, and as the apparatus was costly and imperfectly managed, the expense in the first instance, the trouble, and the noxious smell, presented great obstacles to the introduction of that mode of illumination. The popular notion, also, that streams of flame were rushing along the pipes produced an impression that gas lighting must be very dangerous, which it required time to disprove. It was not, therefore, till several years after the advantages and economy of gas had been practically established, that a public

*It is stated in Mr. Clegg's "Treatise on Coal Gas," that Mr. Clegg, sen., lighted the cotton mill of Mr. Henry Lodge, at Sowerby Bridge, near Halifax, a fortnight before the mill of Messrs. Phillips and Lee was so lighted. A friendly spirit of emulation is said to have existed between Mr. Murdoch and Mr. Clegg in lighting those two mills with gas, each one endeavouring to complete the work before the other.

company was formed for laying down pipes to light the streets, and to convey the gas into houses for lighting shops.

The person to whom the world is chiefly indebted for the practical application of gas lighting is Mr. Winsor, who had been a merchant in London. Being very sanguine as to the advantages to be derived from gas lighting, and possessing an ardent temperament which no opposition could quench, he undertook to introduce it to public notice, and to form a company for lighting the whole of England with gas. He hired the old Lyceum Theatre, which he lighted with coal gas, and he there delivered lectures and exhibited experiments to show the benefits that would arise from the universal use of gas light, and coke fuel. He published an extravagant prospectus of a company to be formed, with the following title:—"A National Light and Heat Company, for providing our streets and houses with light and heat, on similar principles as they are now (1816) supplied with water. Demonstrated by the patentee at No. 97, Pall Mall, where it is proved, by positive experiments and decisive calculation, that the destruction of smoke would open unto the empire of Great Britain new and unparalleled sources of inexhaustible wealth at this most portentous crisis of Europe. The serious perusal of this publication, and an attentive observation of the decisive experiments, will carry conviction to every mind."

In this prospectus Mr. Winsor attempted to make it appear that by adopting his plan there would be "a grand balance of profit for the whole realm of £115,000,000," and each shareholder of the company was promised, "at the lowest calculation, £570 for every £5 deposit." He entertained the notion of making the use of gas and coke compulsory, by levying a tax on all who obstinately refused to adopt what would be so much to their own advantage. This tax, he said, "cannot be oppressive in the least, because it falls on the obstinate only, who shall resist the use of a far superior, cheaper, and safer fuel." Not content with the language of prose, Mr. Winsor vented his thoughts and feelings in numerous poetical effusions. The flights of his Muse, however, were not into the regions of sublimity, as may be perceived by the following specimen:—

"Must Britons be condemned for ever to wallow
In filthy soot, noxious smoke, train oil, and tallow,
And their poisonous fumes for ever to swallow?
For with sparky soot, snuffs and vapours, men have constant strife,—
Those who are not burned to death are smothered during life."

Mr. Winsor's absurd statements—in the truth of which he potently believed—and the wild, random manner of making them known, excited much ridicule and opposition. Among his opponents was Mr. Nicholson, the editor of the *Chemical Review*, who not only challenged Mr. Winsor's estimates, but the validity of his patent, on the ground that Mr. Murdoch was the original inventor. Mr. Winsor's plans and calculations were burlesqued in a cleverly written "Heroic Poem," published in a quarto volume, which, whilst professing to extol the virtues of gas and coke, quizzed its hero most unmercifully. The poem concluded with this address:—

"And when, ah, Winsor! —distant be the day!—
Life's flame no longer shall ignite thy clay,
Thy phosphur nature, active still, and bright,
Above us shall diffuse *post obit* light.
Perhaps, translated to another sphere,
Thy spirit—like thy light, refined and clear—
Ballooned with purest hydrogen, shall rise,
And add a PATENT PLANET to the skies.
Then some sage Sidrophel, with Herschel eye,
The bright *Winsorium Sidus* shall descry;
The *Vox Stellarum* shall record thy name,
And thine outlive another Winsor's fame."

"Though we may smile at Mr. Winsor's extravagant plans and calculations," observes the *Journal of Gas Lighting*, "we cannot but admire the enthusiasm with which he pursued his object, and ultimately succeeded in establishing the first gas company. The lighting of Pall Mall with gas, in the spring of 1807, gave increased stimulus to the project, and application was made to Parliament to carry it into effect. The bill was opposed by Mr. Murdoch and thrown out; but in the following year (1810) the application was successfully renewed. The scheme, however, as sanctioned by Parliament, was sadly shorn of its magnificent proportions; and, instead of a 'Grand National Light and Heat Company, for Lighting and Heating the Whole Kingdom,' its operations were limited to London, Westminster, and Southwark; nor were any special taxes imposed on those who should obstinately refuse to use the light and burn the coke. The Chartered Gas Company, established by Mr. Winsor's persevering efforts, has served as the guiding star to all other gas companies in the world."

The illuminating property of coal gas depends on the quantity of carbon it contains. Pure hydrogen gas burns with a pale blue flame that gives scarcely any light, though it possesses intense heating power. If, however, minute particles of a solid body—powdered charcoal, for instance—be thrown into the flame, they become white-hot, and the incandescence of those solid particles produces a brilliant light. The same effect is caused by the combustion of the carburretted hydrogen gas, and in a more perfect manner. That gas contains a large portion of carbon in a state of vapour, and when a light is applied to a jet of the gas the hydrogen immediately inflames, the carbon is deposited in the flame, and the minute particles into which it is disseminated become highly heated and ignite.

There are two distinct states of carbonization in illuminating gas. The commoner kind—the ordinary coal gas—consists of two measures of hydrogen mixed with one measure of carbon vapour. The specific gravity of such gas is about one-half that of atmospheric air, and it is eight times heavier than pure hydrogen.* The best kind of gas for illumination is

* The facility with which a supply of carburretted hydrogen gas can be obtained from gas works, induces aeronauts to fill their balloons with it rather than be at the trouble and expense of making hydrogen for the purpose; but the ascending power of the balloon is thereby greatly diminished.

obtained from the distillation of oil. It is called olefiant gas, and contains equal measures of hydrogen gas and carbon vapour; its specific gravity is 0.985, being about fifteen times heavier than pure hydrogen.

The *rationale* of the process of making coal gas consists in expelling the volatile matters from the coal by heat, in closed vessels or retorts, and then separating the gas and purifying it on its passage from the retort to the gas-holder, where it is stored for use.

The retorts are usually made of cast iron, and are about 7 feet long, 14 inches in depth, and the same in width; the shape being that of an arch. The retorts will hold two hundredweight of coal each, but they are never filled, because during the process of distillation the carbonaceous part of the coal expands, and occupies more space than the original quantity, the proportion of expansion being as one and a quarter to one. There is a large aperture for the admission of coal and the extraction of coke, which aperture is covered with a lid, and screwed to make it air-tight. A tube is inserted into the mouth of the retort, to carry off the products of the distillation. That tube rises about twelve feet, and then dips into a large horizontal pipe, one foot in diameter, called the hydraulic main, in which are collected the tar and ammoniacal liquor that distil from the coal. Ten or fourteen retorts are usually set back to back in brickwork, to be heated by one fire; but the plan has been recently introduced of employing long clay retorts, which are charged at both ends. These are found to possess considerable advantage over iron, not only from their lower price, but from the facility with which the carbonaceous deposit is removed, and the full charges worked off. Coke is generally burned in the furnaces, and the heat is continually maintained so as to keep the retorts red-hot.

As atmospheric air cannot gain access to the coal in the retorts, the gases expelled do not inflame, nor can the parts that are not volatile be consumed without a supply of air. It is entirely a process of distillation, in which all the products can be collected. The volatile parts are carried off by the pipe, and the solid carbonaceous matter, or coke, is left in the retort.

The first effect of heat on coal, after it is put into the retort, is to expel the moisture, which, in combination with the air, issues in the form of steam. Tar then distils, with some portions of gas, consisting of hydrogen and ammonia. When the retort has attained a bright cherry-red heat, the disengagement of the carburretted hydrogen is most active; and it is found that the more quickly the coal is heated, the greater is the quantity of illuminating gas generated.

The production of coal gas, and the development of its properties at different stages of distillation, may be readily shown by means of a common tobacco pipe. Fill the bowl of the pipe with small pieces of coal, cover it over with a lump of clay, and then put it into a hot fire, with the stalk of the pipe projecting through the bars. Presently steam will be seen to issue from the pipe, and afterwards smoke, and, if a light be applied, a jet of flame will issue forth, the brilliancy of which will increase as the bowl of the pipe becomes more heated, until the best part of the gas has been distilled from the coal.

The gas is mingled with various volatile products as it issues from the retort, and requires to be purified before it is fitted for illumination. The most abundant matter that passes over with it is tar. The vapour of that substance, however, condenses when cooled, and it may thus be separated from the gas very effectually. For that purpose the gas, after having deposited a large portion of the tar in the hydraulic main, is made to traverse through a number of vertical pipes, and in passing through them a further quantity of tar, accompanied by ammoniacal liquor, is deposited, and collected in a reservoir at the bottom. The next process is the purification of the gas from carbonic acid and sulphuretted hydrogen. This is commonly done by passing it through water and lime; the combination of the carbonic acid with the lime being facilitated by agitation. The method of purifying by lime was introduced by Mr. Clegg; and by a later process, oxide of iron is used to absorb the sulphuretted hydrogen. The gas, when purified, is conveyed to the gas-holder, whence it is forced by pressure into the mains and pipes.

An apparatus for generating coal gas on a small scale for private establishments, remote from sources of ordinary supply, is represented in the accompanying woodcut. The retort, A, is fitted in a small furnace. The coal is put in at F, and the products of distillation pass through the bent pipe, E. The more liquid portions of the tar pass at once through the tube, B, into the receiver, G; and as the gas passes along the bent tube, C, it becomes cooled, and a further deposit of tar and ammoniacal liquor is made.

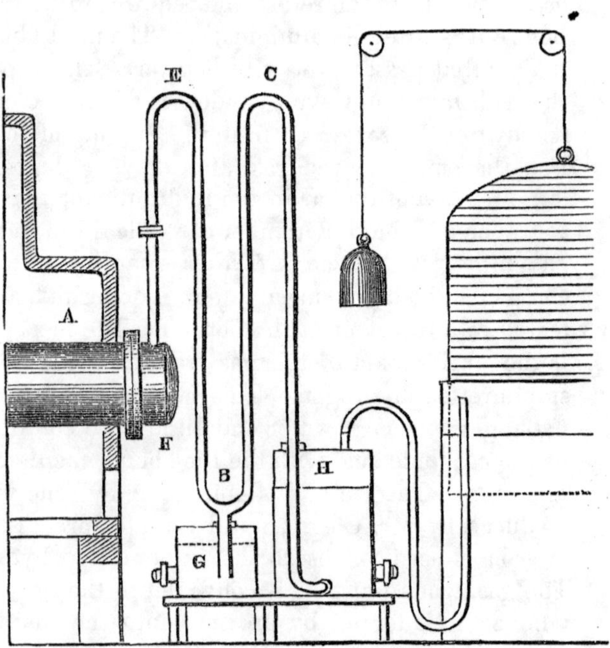

The gas is then conveyed along another tube into the purifier, H, filled with lime and water, and it thence passes into the gas-holder. Tubes are inserted into the latter for conveying the gas to the burners.

The quantity and the quality of the gas yielded by coal differ materially according to the kind employed. One ton of good Newcastle coal will yield 9,500 cubic feet of gas, which, when burnt in the best manner, gives a light equal to that of 422 lbs. of spermaceti candles. One ton of Wigan cannel coal yields 10,000 cubic feet, and gives a light equal to 747 lbs. of spermaceti candles.* The price, in London, of good gas from Newcastle coal, is 4s. 6d. per thousand cubic feet, which gives a light equal to 74½ lbs. of spermaceti, and equal to 89 lbs. of mould candles; therefore, when the latter are 8d. a pound, the burning of gas is twelve times more economical than the burning of candles. In Liverpool, gas from cannel coal is supplied at the low price of 3s. 9d. per thousand feet; and that gas gives at least one-third more light than the ordinary London gas.

The cleanliness of gas, as compared with candles or oil, is a further recommendation; and for the purpose of lighting streets, shops, factories, public buildings, and halls, it presents important advantages; but it is not well adapted for small sitting rooms, because the heat of the flame makes it unpleasant and injurious to the eyes when near, and, unless very pure, it deteriorates the air of closed apartments. In many parts of the country, however, where coals are cheap, and the price of gas is consequently less than in Lon-

* *Journal of Gas Lighting*, vol. ii.

don, it is introduced into every room of nearly all private houses.

The best kind of gas made from mineral substances is produced by the distillation of a bituminous shale, called Boghead coal, which was discovered a few years since in Scotland. One ton of this material yields 15,000 cubic feet of gas, which is equal in illuminating power to 1,930 lbs. of sperm candles. Boghead coal is now commonly used for mixing its gas with that of inferior quality, to bring up the illuminating power to the required standard.

Olefiant gas, made from oil, burns with a brighter and purer light than common coal gas, but it is more costly. It is made nearly in the same manner, by distillation in retorts; the principal difference consisting in the degree and regulation of the temperature. A dull red heat is the best, and in order to keep the oil exposed to the action of an invariable heat, it is admitted gradually into the retorts, into which pieces of brick or coke are inserted to increase the heating surface. One pound of common oil yields about 15 feet of olefiant gas. The same kind of gas may also be obtained in smaller quantities by the distillation of tar, rosin, or pitch. Twelve cubic feet of gas may be obtained from one pound of tar, and ten from the same weight of rosin.

The brilliancy of gas-light depends, in some measure, on the kind of burner employed. To obtain a steady light, an argand burner is usually adopted; the gas being allowed to escape through a number of minute holes pierced in a hollow ring of metal, which admits a current of air through the middle. To increase the supply of air, the burner is covered with a glass chimney, which, if not too long, adds to the brilliancy of the flame; but a very long chimney produces so strong a current of air, as to cool the flame, and diminish the light. A plan is sometimes adopted of placing a small metal disc a short distance above the jets, so as to spread the flame. By this means the brightness is increased, by exposing the flame more directly to the current of air; and the metal disc, by becoming heated, also tends to aid the combustion of the carbon.

One of the problems to be solved on the original formation of gas works was the size of pipes, and the amount of pressure required to force the gas to the various burners. It was at first supposed that the friction against the pipes would oppose so much resistance to the passage of the gas, that it could not be transmitted to great distances. It was found, however, that the perpendicular pressure of a few inches of water was quite sufficient to force the gas through the mains and small pipes of an extensive range of streets. A bold attempt was made at Birmingham, in 1826, to bring gas from the collieries, at a distance of ten miles from the town. The plan was laughed at by many as impracticable, but it was attended with complete success. The gas being made near the mouth of the coal-pit, the cost of conveyance was saved by the additional outlay in the first instance. It must be observed, however, that it is extremely difficult in practice to avoid the escape of gas at the junctions of the pipes; and by increasing the length of the gas mains, the greater will be the leakage. The loss from this cause, in some gas works, exceeds 20 per cent. of the gas manufactured.

The volume of gas discharged from a pipe is directly proportional to the square of its diameter, and inversely as the square of its length. Thus, if a pipe required to discharge 250 cubic feet of gas in an hour, at a distance of 200 feet, must have an internal diameter of 1 inch; to discharge 2,000 feet in an hour, at a distance of 1,000 feet, would require a diameter of 4·47 inches. The same quantity discharged at double the distance would require a pipe 5·32 inches in diameter; at a distance of 4,000 feet the diameter must be increased to 6·13 inches; and at a distance of 6,000 feet the diameter should be 7 inches.

On the first introduction of gas-light, the companies who supplied it charged a fixed sum for each burner of a given size. This mode of charging was, however, very unsatisfactory, for the size of the burner is a very uncertain indication of the quantity of gas consumed. Persons using gas desired to pay for the quantity they actually burned; and to enable them to do this, a special contrivance was invented by Mr. Clegg, the engineer of the Chartered Gas Company, called a gas-meter. That instrument measures, with sufficient accuracy for practical purposes, the volume of gas that passes through it to the burners, and thus each consumer of gas now pays only for the number of cubic feet consumed.

The accompanying diagrams represent sections of a gas meter, as seen in front and edgewise. The outer case of the instrument, which is a flat cylinder made of sheet iron, is indicated by the letters *c*, *c*. Inside it there revolves another cylinder, made also of thin sheet iron, and divided into four compartments, marked *d*, *d*, *d*, *d*. This interior cylinder readily revolves on an axis, *g*, *g*, shown in the section of the instrument as seen edgewise. The gas enters from the street pipe through the opening, *a*, and it is forced out to the burners through the pipe, *b*, the latter being seen in the narrow section only. In that

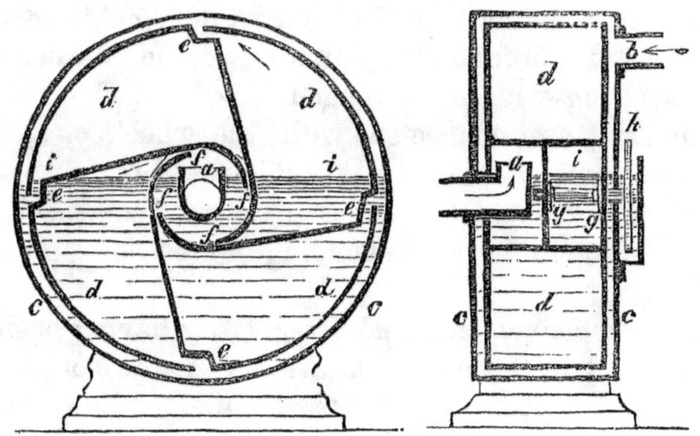

diagram, also, there is shown a cog-wheel, *h*, fixed on to the axis, and a small outer case, in which that wheel rotates. Water is poured into that external case until the gas-meter is rather more than half filled, the level of the water being shown at *i*.

The action of the instrument will be readily understood by examing the two sections. The gas, on entering the tube, *a*, presses against the upper surface of the compartment that happens to be then above it, and tends to turn the inner cylinder round. This pressure forces the gas through the opening, *b*, to the burner; and as the compartment then in communication with that opening is emptied of the gas it contains, in the direction of the arrow, it is gradually forced under the level of the water, and the other compartment, which has in the meantime been filling with gas, continues the supply. Thus, supposing each division of the inner revolving cylinder to hold 108 cubic inches, a complete revolution would indicate that the fourth part of a cubic foot had passed through the pipe, *b*, to the burners. Several cog-wheels, arranged like clock-work mechanism, are connected with the wheel, *g*, and by this means the number of cubic feet of gas consumed is indicated by hands fixed to the wheels, and pointing to the corresponding figures on a series of dials.

Some inconvenience and irregularity having been experienced in the use of the wet meter, the correctness of which, it is evident, may be affected by variations in the height of the water level, dry meters have been constructed for measuring gas, by causing it to pass through a small expanding chamber, similar in principle to a pair of bellows. The objection to these instruments is that the leather, or other flexible substance that forms the sides of the expanding chambers, becomes rigid by use, and the valves are liable to get out of order; but in the last improvement of the instrument, by Mr. Croll, these objections are stated to be effectually removed.

Numerous attempts have been made to produce illuminating gas from other substances than coal, but without advantage. The plan that promised the most success was the production of hydrogen gas by the decomposition of water, which was passed over heated coke in retorts, and by that means the oxygen of the water, combined with the incandescent coke and the hydrogen, was set free. The gas thus collected possessed little illuminating power, but it was afterwards mixed with the rich gas from cannel coal, and raised to the requisite illuminating standard. It was found, however, in practice, that the compound gas thus formed was more costly than ordinary coal gas, and the plan has been discontinued. Another method of giving illuminating power to water gas was to surround the flame with platinum gauze, which was rendered incandescent by the heat, and became highly luminous. But it required twice the quantity of gas burned in this manner to produce a light equal to that of carburretted hydrogen, and the combustion of so much hydrogen gas produced an amount of vapour and heat that were very unpleasant. That mode of gas illumination, called the "Gillard light," from the name of the inventor, was also found more costly than the ordinary mode of lighting with coal gas, which has now no rival to compete with it in economical illumination.

No Act of Parliament is now required, as

originally proposed by Mr. Winsor, to enforce the burning of coal gas. Its advantages, in point of economy, cleanliness, and even of safety, are sufficiently understood to spread the use of coal gas to every part of the kingdom. In the metropolis alone there are twelve gas companies, who receive for the sale of gas an average of £100,000 per annum each, thus making the sum paid for gas lighting in London £1,200,000, and it has been estimated as high as £2,000,000. Taking the average price to be 4s. 6d. per thousand cubic feet, the quantity of gas consumed amounts to 5,300,000,000 cubic feet; and if we add to that quantity 20 per cent. for leakage through the mains and pipes, the quantity of gas manufactured in the metropolitan gas works is upwards of 6,000,000,000 cubic feet in a year. It may, perhaps, give a clearer notion of this immense quantity to say, that a gas-holder, capable of containing it, would require to be one mile in diameter, and the height of St. Paul's Cathedral. The light produced by burning such a volume of gas would be equal to that of 150,000 tons of mould candles, which would cost £13,000,000. The quantity of coals requisite for the production of the gas manufactured annually in London is upwards of 600.000 tons.

THE ELECTRIC LIGHT.

The Electric Light is the brightest meteor that has flashed across the horizon of promise during the present century. When first exhibited as a means of illumination, about twelve years ago, the splendour of the rays emitted, and the delusive representations of the small cost required to produce such a brilliant light, led the public to believe that the career of gas-lighting was drawing to a close, and that night would be turned into day by this wonderful demonstration of electrical power. The light produced by charcoal points, subjected to the action of a powerful voltaic battery, was, however, no novelty at that time; for as far back as 1810, Sir Humphry Davy was accustomed to exhibit that development of electrical force at the Royal Institution, and it formed a standard experiment in most chemical lectures. But it seems not to have been thought applicable in those days to the purposes of illumination; and when Mr. Staite brought it into notice, and exhibited its effects on the tops of some public buildings, it was considered one of the most wonderful inventions of the age.

Mr. Staite's patent, taken out in 1847, though commonly supposed to be for the Electric Light generally, was limited in its clauses to the construction of a voltaic battery and apparatus, adapted for maintaining constancy, and for giving steadiness to the light. The merely temporary continuance of the *voltaic arc*, as it was formerly called, seemed indeed to preclude the possibility of its adoption as a means of illumination ; it was therefore a great point gained to give stability and constancy to the light. The difficulty of accomplishing this will be perceived when it is known that the charcoal points, between which the action takes place, are constantly undergoing change, the particles of carbon being transferred from one to the other. There is no actual combustion of the charcoal, in the ordinary meaning of the term ; the action is principally confined to the transfer of the charcoal connected with the positive pole, to that connected with the negative pole of the voltaic battery, a hollow being formed in one, and a pyramidical accumulation of particles in the other. This action was beautifully shown by Professor Faraday at the Royal Institution last year, by projecting the image of the charcoal points on to a screen, by means of the Electric Light itself. The image, magnified by the lenses of the electric lamp, could thus be distinctly seen without being too brilliant to dazzle the eyes. The particles of carbon, heated to whiteness, were perceived to be in active motion, and the piling up of the pyramid in one, and the hollow produced in the other, were continually varying the distances between them, and thus tending to cause unsteadiness in the light.

Numerous contrivances have been adopted for the purpose of keeping the points at exactly the same distance, as the want of stability was supposed to be the only obstacle to the adopton of the Electric Light. These contrivances have so far succeeded, that a tolerably steady light can be maintained for some time, but under the most careful management the points occasionally approach too near or are too far apart to maintain an equable light.

Among other inventions to increase the steadiness of the light is one that was patented in 1856, by Mr. Way, in which mercury is substituted for charcoal, but the steadiness of light to be thus acquired must be attained with a great loss of illuminating power, and the vapour arising from the combustion of the mercury would be extremely injurious to health.

Mr. Hearder, of Plymouth, has produced more brilliant effects with the Electric Light than any other person. Some remarkable exhibitions of the power of the light were made by him, in April, 1849, from the top of the Devonport Column, and several scientific gentlemen undertook to make observations at different localities to a distance of five miles. At Tremeton Castle, on the banks of the Tamar, a distance of nearly 3½ miles; the light cast a strong shadow, and writing could be distinctly read by it. The space illuminated was at least three quarters of a mile broad. To aid the effect, a reflector was employed, and when the rays were directed to the clouds, they had the appearance of a huge comet, the reflector being the nucleus. The intensity of the light was ascertained to be equal to that of 301,400 mould candles of six to the pound, whilst the light of the Breakwater Lighthouse was equal to only 150 candles. At a distance of five miles the light was sufficiently powerful to enable persons to read a book.

The battery employed by Mr. Hearder in these brilliant experiments consisted of 80 cells of a Maynooth battery, 4 inches square, and the carbon cylinders between which the light appeared were formed of powdered coke, mixed with tar, and rammed into a tube three quarters of an inch in diameter. When these cylinders are about a quarter of an inch apart, the Electric Light appears at the end of each for the space of more than half an inch. The light, during the experiments at Plymouth, was maintained for three hours, and the materials employed amounted to one pound and a-half of zinc, 114 fluid ounces of sulphuric acid, the same quantity of nitric acid, and six pounds of muriate of ammonia. *

The most serious practical objection to the intro-

* Mr. Hearder, of Plymouth, affords a remarkable instance of the successful pursuit of science under difficulties. He lost his sight in his youth by an accidental explosion during some chemical experiments, but instead of being disheartened by that calamity, he has continued to pursue his investigations with unabated vigour, and has succeeded in throwing much light on many of the recondite properties of electricity, by admirably contrived experiments, which were conducted with unremitting perseverance at great expense. He has been in the habit of delivering lectures at the Plymouth Institution, and other Institutions in Devon and Cornwall; and those who witness the skilful manipulation of his experiments can scarcely suppose that he is blind.

duction of the Electric Light, as a means of general illumination, is its expense. When the project was first brought into notice, attempts were made to show that the battery power required might be obtained at little cost, and in this respect some deceptions were practised not creditable to the parties engaged in promoting the scheme. It has been proved by Mr. Grove that the cost of ordinary batteries necessary to maintain the light in full brilliancy would greatly exceed the price of an equal light from gas.

A plan was patented for generating the required voltaic power, free from cost, by applying the residual sulphate of zinc as paint, and an Electric Power and Light Company was formed to carry out the project. But the plan failed, and the affairs of the company have recently been "wound up."

Until some cheaper mode of generating electricity than is at present known be invented, there is no hope of the Electric Light becoming generally available, but there are special circumstances in which it may be applied with advantage. It is peculiarly applicable for lighthouses, as its rays would penetrate through a foggy atmosphere that would obscure the light of ordinary flames, and in such cases the extra cost should not operate as an obstacle to its use.

INSTANTANEOUS LIGHTS.

Those who are not old enough to remember the time when flint-and-steel were the implements employed to obtain a light, can have no sufficient appreciation of the great convenience of "Lucifer" matches. In those "good old times," it was a regular household care to provide a sufficiency of tinder, to see that it was kept dry, and that there was a proper flint "with fire in it." The striking of a light, when the tinder-box was adequately supplied, was no mean accomplishment; and the unskilful hand, operating in the dark, would either get no sparks at all, or send them in a wrong direction, and not unfrequently strike the skin off the knuckles, in the vain endeavour to set light to the tinder. Or if the tinder were damp, the sparks would fall upon it without igniting, and minutes would be spent in holding a pointed brimstone match to the delusive spark, and blowing at it without effect. Sometimes the incautious operator, tired with his fruitless efforts, would sprinkle gunpowder over the tinder, to make it take fire more readily, and whilst puffing at a long-desired spark, the gunpowder would explode in his face and nearly blind him. Such were some of the annoyances, attended by loss of time, that were experienced in obtaining thè same result that is now produced instantaneously, and much more effectively, by merely rubbing the match against any rough surface.

Several attempts had, indeed, been made many years ago to supplant the flint-and-steel and tinder-box, and some of the plans adopted so closely approach the matches now in use, that we wonder the inventors did not succeed long since in contriving the very facile means of striking a light that we now enjoy. Phosphorus and brimstone matches were first employed for the purpose. The phosphorus was contained in a bottle placed within a tin case, which also held the pointed brimstone matches and a piece of cork. The match was dipped into the phosphorus bottle, and then rubbed on the cork; and the friction excited sufficient heat to inflame the small quantity of phosphorus adhering to the match and, to set fire to the sulphur. These phosphorus boxes answered the purpose very well, but the apprehended danger of using so inflammable a substance prevented their coming into general use; and they were much more costly than a tinder-box.

In the next advance, if it may be so called, in the invention of instantaneous light-producers, phosphorus was altogether discarded, and a mixture of chlorate of potass, then called oxymuriate of potass, and sugar was employed. Those substances, when combined, inflame explosively in contact with sulphuric acid. In applying them for the purpose of obtaining instantaneous light, they were mixed together in an adhesive menstruum, into which the ends of small rectangular matches were dipped. These matches very nearly resembled the "Lucifers" of the present day. To ignite them, a small bottle containing sulphuric acid and asbestos was provided, and they were arranged together in an ornamental taper-stand for the chimney-piece. This apparatus was not received with much favour, partly on account of injury done by a careless use of the sulphuric acid, partly because it failed to act when the acid had absorbed moisture from the atmosphere, but principally because of its cost.

To obviate the objection arising from the use of sulphuric acid in open bottles, an ingenious contrivance was adopted, by which each match contained its own reservoir of acid sufficient for igniting the inflammable compound. Small glass globules, containing sulphuric acid, were introduced into the composition of chlorate of potass and sugar, which, when broken, set fire to the mixture and lighted the match. These instantaneous lights, which were called *Prometheans*, were more ingenious than useful, because the trouble of manufacture rendered them expensive, and the sulphuric acid was more likely to injure furniture in that form than when a bottle with asbestos was used. The Prometheans, however, possessed the advantage of portability, and for occasional purposes they were convenient. In some of the forms in which the Prometheans were manufactured, the glass globule of acid, surrounded by its inflammable compound, was attached to the end of a small stick of sealing-wax, sufficiently large to seal a letter; but this refinement in instantaneous lights was not much patronized.

Notwithstanding these ingenious attempts to produce light by chemical action, the flint-and-steel retained possession of the field until a match was made that ignited by friction alone. The first kind of friction match was invented in 1832. It consisted of a thin splinter of dried wood, the top of which was dipped. in a mixture of one part of chlorate of potass, two of sulphide of antimony, and one of gum. To ignite the match it was necessary to draw it briskly through sand-paper. These matches required some address to light them, because much more friction was required than is sufficient to light Lucifers.

The next improvement was the "Congreve" match, in which recourse was had to the materials previously used, separately, for obtaining instantaneous lights. Congreve matches were composed of an emulsion of phosphorus mixed with chlorate of potass, into which the matches, previously tipped with sulphur, were dipped. These matches were of the same size and form as the Lucifers now in general use, and they ignited readily by friction on sand-paper or other rough surface. Their explosive noise on inflammation, which gave them their name, was the only apparent difference between Congreves and Lucifers, and their introduction completely supplanted the flint-and-steel.

The noiseless match, or Lucifer, has, in its turn,

driven the Congreve almost out of use, though for practical purposes the latter was as effective, and it was less dangerous. The Lucifer matches depend altogether on phosphorus for their inflammability. Their composition is an emulsion of phosphorus with glue, nitre, and some colouring matters. The sulphur matches, after having been tipped with that composition, are exposed in a warm room until a sufficient quantity of the phosphorus is evaporated by slow combustion, to leave a film of glue on the surface to protect the remainder from the action of the atmosphere. The usual proportions for the compound are, phosphorus four parts, nitre ten, glue six, red ochre five, and smalt two. The principle on which the action of Lucifer matches depends, is the strong affinity of phosphorus for oxygen, of which the nitre with which it is mixed contains an abundant supply; and by drawing the match across sand-paper, sufficient heat is excited by the friction to ignite the phosphorus, and the nitre supplies the oxygen to maintain rapid combustion.

The manufacture of Lucifer matches is conducted on a very large scale in this country and on the Continent. It requires several ship loads of wood to supply the requirements of Lucifer-match makers; and ingenious contrivances have been patented for cutting it up into splints of the proper size. For that purpose, after the wood has been reduced to the required lengths by circular saws, it is cut up into splints by a number of lancet points, separated from each other as far apart as the thickness of a match, which pass over the wood and divide it with great rapidity. The splints are collected into bundles of one thousand, and each end having been dipped into melted sulphur, they are divided in the middle by a circular saw.

The Reports of the Juries of the Great Exhibition supply a variety of statistical details respecting the manufacture of chemical matches, from which it appears that the quantity made in Austria, in 1849, amounted to 50,000 cwt.; and that in France, in 1850, the phosphorus consumed in the manufacture of matches, amounted yearly to 590 cwt.; and the consumption has rapidly increased since that time. In this country, it is calculated that eight tons of phosphorus are yearly used in making matches, the number of which is stated to be 40,000,000 a day. Large quantities are also imported from Germany, where they are manufactured so cheaply, that fifty boxes each containing 100 matches, are sold for fourpence.

The latest improvement in chemical matches is the "Vesta," which consists of small wax, or stearine tapers, with an igniting composition at the end, consisting of chlorate of potass and phosphorus. These Instantaneous Lights are made without sulphur, consequently the disagreeable smell of the common Lucifer is avoided. The convenience of smokers has also been consulted in the manufacture of Instantaneous Lights. The fusees, now so frequently used for lighting cigars, are composed of thin card-board cut half through, steeped in nitre and with a small quantity of phosphorus; and the tearing of the paper across produces sufficient heat to ignite the inflammable card.

Thousands of persons, principally children, are now employed in the manufacture of chemical matches. The occupation, as at present conducted, is very unhealthy, for the fumes of the phosphorus produce a disease of a remarkable kind in the jawbone, which often proves fatal. No cure has yet been found for this peculiar disease, occasioned by the phosphorus in the state in which it is commonly used. A preparation of that substance has, however, been made which may be used without injury, and which possesses the advantage also of being less dangerously inflammable; but as the red *amorphous phosphorus*, as it is called, is rather more costly, the manufacturers of Lucifer matches object to use it.

PAPER MAKING MACHINERY.

CHEAP literature and the large development of newspapers are principally attributable to the improvements in Paper Making, by the aid of machinery.

In the former modes of making paper, the workman held in his hands a square frame covered with wires, which he dipped into the prepared cotton or linen pulp, which was kept in suspension by being agitated in water, and taking up a quantity sufficient to cover the frame, he moved the pulp about horizontally, to spread it evenly over the surface of the wires. Another workman transferred the layer of pulp on to felt, and in this manner one sheet was laid upon another, with felt between each. They were next subjected to great pressure, for the purpose of making the fibrous particles cohere sufficiently to form sheets of paper. The felts were then removed, and the sheets were piled upon one another and again pressed, after which they were dried, sized, and finished.

Paper Making, by that process, was a slow operation. The thickness and evenness of the sheets depended altogether on the judgment and skill of the workman, and their size was necessarily limited by the dimensions of the frame. By the improved methods, nearly all the work is done by machinery. The soft fibrous pulp, which is to be converted into paper, enters the machine at one end, and in the course of two minutes it is delivered at the other end of the machine in a continuous sheet, that may extend for miles. By supplemental contrivances the paper is cut into sheets, piled together, and presented in a salable form.

The world is indebted to a Frenchman, named Louis Robert, for the invention of the first machine for making paper. He was a workman in M. Didot's paper mill, at Essones, and for his contrivance of a method for making continuous paper, he obtained from the French Government, in 1799, the sum of 8,000 francs and a patent for the manufacture of the machines. The political agitation in France at that period prevented much progress from being made with the invention, but after the Peace of Amiens, in 1802, M. Didot, jun. came to this country, accompanied by his brother-in-law, Mr. Gamble, for the purpose of making arrangements to carry it into effect. They induced Messrs. H. and S. Fourdrinier to engage with them in bringing the machinery to perfection, and patents obtained in this country by Mr. Gamble were assigned to them in 1804.

The engineering establishment of Mr. Hall, at Dartford, in Kent, was selected as best adapted for the purpose of making the machinery and for carrying the plans into operation. Mr. Bryan Donkin, who was engaged in the manufactory, principally assisted in bringing the machinery to perfection. The difficulties attending the completion of all the parts, to get them to work effectually, and the obstruction encountered in introducing the machine-made paper, rendered the enterprise a ruinous speculation to those who first engaged in it. Messrs. Fourdrinier having expended £60,000 in perfecting the machine.

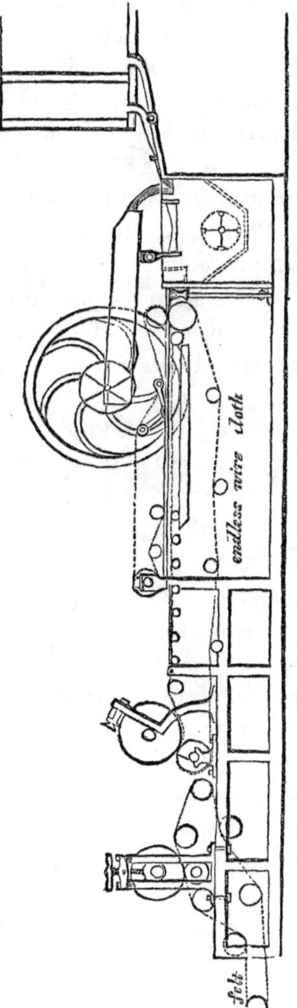

The apparatus, of which a representation is given in the annexed woodcut, was very complicated, but the essential parts may be readily understood.

The rags from which the paper is made undergo a variety of processes before they are properly reduced into a state of pulp. They are sorted, dusted, boiled, and torn into pieces by passing through cutting rollers; they are then bleached and again submitted to the grinding action of rollers, which reduce them into a state of fine pulp, resembling milk in appearance. The pulp thus prepared is placed in a large vat, where it is kept constantly agitated, to prevent the more solid parts from being deposited. From the vat the pulp is discharged into a cistern, over the edge of which it flows in a continuous stream upon an endless wire cloth, the meshes of which are so fine that there are as many as 6,000 holes in a square inch.

The wire gauze, on to which the pulp is poured, is about 4 feet wide, and 25 feet long, and it is kept constantly moving onwards, by rollers at each end, over which it passes. The gauze is stretched out perfectly level, and the pulp is prevented from flowing over the edges by straps on each side, which limit the width of the paper. As the endless wire cloth moves along, an agitating motion is given to it, by which means the pulp is spread evenly over the surface; the water is also drained off through the interstices of the gauze, and this part of the process is expedited in the improved machines by producing a partial vacuum underneath. Before the sheet of pulp has arrived at the farther extremity of the wire cloth, it passes between two cylinders, the under one of which is of metal, covered with felt, and the upper one of wood. A slight pressure given to the pulp in passing between those cylinders imparts sufficient tenacity to it to enable it to be transferred from the wire gauze on to an endless web of felt, by means of a slice that clears the pulp from the wire gauze, and deposits it on the felt. The latter is kept moving at exactly the same speed as the wire gauze, otherwise there would be either a rent or a fold on the sheet. The paper, still in a very wet state, is carried between cast iron rollers, and its fibres are forcibly pressed together, which operation squeezes out the water, and so far gives tenacity to the pulp that it may be handled without tearing. The sheet then passes on to other rollers, by which it is further compressed, and its surface smoothened. The paper is, however, still damp, and requires to be dried. This is done by passing it over large metal cylinders, heated by steam. The process of making the paper is then completed, and the continuous sheet may be wound upon a reel to any length; but it is now usual to cut it up into sheets as soon as it leaves the drying cylinders.

The wire cloth moves at the rate of from 25 to 40 feet per minute, and such a machine would consequently make at least 10 yards of paper in that time, which is equal to a mile in three hours. The width of the paper is usually about 4½ feet, therefore each machine will make 10,450 square yards of paper in twelve hours; and there are upwards of three hundred of such machines at work in this country. The value of the paper thus produced is calculated to exceed two millions sterling.

Numerous improvements have been made in Fourdrinier's original machine, but the principle of its construction remains essentially the same, and it is by this means that most of the paper now used for writing or printing is manufactured. A paper-making machine, on a different principle, has, however, been invented by Mr. Dickinson, and has been carried by him to great perfection. Instead of allowing the pulp to fall on to a flat surface of wire gauze, a polished hollow brass cylinder, perforated with holes and covered with wire cloth, revolves in contact with the prepared pulp, and a partial vacuum being produced within the cylinder, the pulp adheres to the gauze, and its fibres cohere sufficiently, before the cylinder has completed a revolution, to be turned off on to another cylinder covered with felt, on which it is subjected to pressure by rollers, and is thence delivered to the drying cylinders.

Mr. Dickinson afterwards obtained a patent, in 1855, for making a union paper, consisting of a thin sheet of that made by his own machine, and a similar sheet made by a Fourdrinier machine united together. For this purpose the two sheets were brought together, as they passed from the machines, whilst still wet and in an unfinished state, and were pressed together between rollers, by which means they were completely incorporated. The object of this contrivance was to combine, in a single sheet, the different kinds of surface which paper made by those two modes of manufacture present. It is also employed economically for engravings, to give a fine surface to a thick sheet of coarser material. The threads in postage envelopes and in bankers' cheques, are introduced by this process of plating two surfaces together.

The greatly increased consumption of paper threatened to exhaust the supply of the raw material, notwithstanding the large import from abroad and the enormous supply derived from the waste of the cotton mills, which, when mixed with rags, produces good paper. The quantity of old rags, old junk, and other fibrous materials imported for the purpose of making paper, in 1850, is stated in the Jury Reports of the Great Exhibition to have amounted to 8,124 tons. This large importation, added to the stock of rags supplied by the country itself, was, however, inadequate to meet the consumption, and search was anxiously made for other fibrous substances that could be converted into paper;—peat, cocoa-nut fibre, grass, straw, and even wood have been used for the purpose. Of those substances, straw has been most successfully applied, and straw paper—excellent to write upon, though not bright in colour—is now made at very low prices. The straw is first cut up into short lengths, of about half an inch, by a chaff-cutting machine, and after undergoing various processes of trituration and bleaching, it is reduced into a pulp, sufficiently adhesive to make a strong paper.

The plan of drying the paper as it leaves the rollers of the machine, was introduced by Mr. Crompton in 1820, and that gentleman was also the first to introduce a machine for cutting the paper into sheets as soon as it is dried. The first invention of the kind was patented by Mr. Crompton, in conjunction with Mr. Miller and Professor Cowper, in 1828. The continuous web of paper was made to pass directly from the drying apparatus to the cutting machine, by which it was first slit into bands of the required width by means of a series of sharp discs of steel, adjustable on two parallel axes. The bands of paper then passed on to shears, placed transversely, that cut it into sheets of any required length, which were laid upon one another, to be divided into quires.

Several other cutting machines have since been invented, the simplest of which is the one patented by Mr. Dickinson, which is represented in the woodcut.

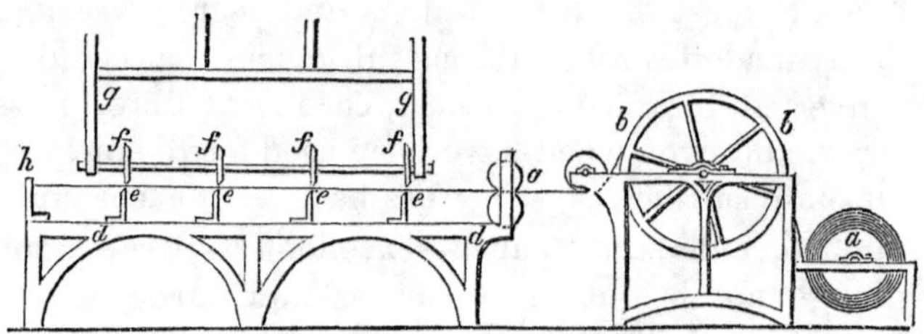

The paper may be taken directly from the drying cylinders or from a reel, as shown in the diagram at *a*. The sheet passes over a large drum and through several guide rollers, till it is carried across the table *a h*, where it is cut lengthwise by knives, as it passes along. A series of chisel-edged cutters are placed at regulated distances beneath the table; and whilst the paper is stretched over it, several circular knives, *f f*, fixed into a swing frame, *g g*, at corresponding distances with the knives beneath, are swung across the sheet, and cut it in the manner of a pair of shears. Other kinds of cutting machines are contrived, by which sheets of writing paper, when collected in quires, are squeezed tightly together, and their edges are smoothly and evenly cut.

We must not conclude this notice of Paper Making Machinery without alluding to the ingenious self-acting mechanisms for making envelopes. In the Great Exhibition of 1851 there were three different machines exhibited in action, each one producing, with great rapidity, those neat coverings for letters, for which the penny postage system has created so great a demand. The paper, cut into the desired form by a separate machine, was piled up on one side of the envelope folder. It was taken, sheet by sheet, and stretched on a small table, on the middle of which there was a trap door, held up by a spring to a level with the rest of the table. A plunger, of the same size as the envelope to be made, pressed the trap down into a recess, and raised the four corners of the paper, the edges of which were then gummed, and small mechanical fingers folded them down. The completed envelope was then thrown out into a basket, or it slided out of the machine on to those before made.

Each of those machines, with a boy as an attendant, will fold 2,700 envelopes in an hour, which is nearly the same number that an experienced workman can fold in a day with a folding stick. Notwithstanding the supplanting of manual labour to so great an extent by these ingenious mechanisms, the effect of increased facility of manufacture has been to give increased employment, and many more persons are now engaged in making envelopes than were so employed before the invention of the machines.

PRINTING MACHINES.

THE associated inventions of paper making and printing have progressed hand in hand together; the increased facility with which paper can be made by machinery having been equalled, if not surpassed, by the rapidity with which it can be printed.

The old wooden printing press, that was in general use at the beginning of the present century, is now an object of curiosity, and a few specimens of it are to be seen, even in country printing offices.

The principal working part of the wooden press consisted of a block of wood, having a perfectly flat and smooth surface, half the size of an ordinary sheet of printing paper, which was brought down upon the types by means of a screw that was turned by a long lever. The types, placed upon a flat stone embedded in a movable table, were inked with large soft balls covered with pelts. The damped paper was put into a frame, at the back of which blankets were placed, and was laid lightly on the inked types. The movable table was then pushed under the block of wood, called the "platten," the long lever was pulled with great strength, and the platten being thus brought forcibly upon the blankets and paper, one-half of the sheet was printed. The lever, on being released, sprang back to its former position, and the table with the types upon it was pushed farther under the platten, which was again pulled down to print the other half of the sheet. The table was then pulled back, and the sheet of paper, printed on one side, was removed. These operations occupied considerable time, and the regular work of two men, with a wooden press, was to print 250 sheets an hour on one side.

This original contrivance for printing was supplanted by the Stanhope press, one of the most admirable arrangements for the advantageous application of the lever that is to be found in the whole range of mechanical contrivances.

The improved printing press, invented by Lord Stanhope, the first of which was completed in 1800, is made altogether of iron. The platten is of the full size of the sheet of paper to be printed, and the work is done at a single pull. The requisite power is obtained by a combination of levers, so adjusted that the platten is brought down rapidly in the first instance, before any pressure is required, and when it comes to bear upon the types, the levers act with the greatest possible mechanical advantage, so that the handle moves through the space of a foot, whilst the platten descends only the twentieth part of an inch. By this means a large sheet of paper can be printed off by a single pull, and with more impression and greater sharpness than by two pulls with a wooden press.

Great as was this improvement in the printing

press, its action was still very slow, compared with the rapidity of printing we are now accustomed to, it being considered quick work, with a small Stanhope press, to print 500 sheets an hour. The author remembers to have seen the *Globe* newspaper printed by an old wooden press in 1820; and, about the same time, the London *Courier*, by a Stanhope press. In order to supply the large demand for the latter paper, it was then customary to print off three pages early in the day, and to set up the types for the fourth page, containing the latest news, three or four times, and to print it at as many separate presses. The pressmen could thus, by great exertion, perfect the printing, when three presses were used, at the rate of 1,500 an hour. The *Times* newspaper, which greatly exceeds the size of the *Courier*, is now printed by a machine at the rate of 13,000 an hour.

The invention of printing machines was preceded by the manufacture of inking rollers, to supersede the pelt balls for distributing the ink over the types. Earl Stanhope had endeavoured in vain to construct inking rollers, for which purpose he tried skins and pelts of various kinds, but the seam proved an obstacle that he could not overcome. In 1808, a "new elastic composition ball for printing," which consisted principally of treacle and glue, to serve as a substitute for pelts, was invented by Mr. Edward Dyas, a man of great original genius, the parish clerk of Madeley, in Shropshire. These balls were first introduced into the extensive printing office of the late Mr. Edward Houlston, of Wellington, where they were for some time exclusively used, and that printing-office consequently became celebrated for the excellence of its work. A similar composition was some years afterwards cast in the form of rollers, upon a hollow core of wood, by the late Mr. Harrild; and these rollers have proved a far more cleanly and more expeditious mode of inking the types than the balls. These inking rollers supplied an essential want in the working of Printing Machines.

The invention of Printing Machines underwent many changes before it was brought to a practical form. Such a machine was first projected in 1790, by Mr. Nicholson, who proposed to place the types and paper upon cylinders, and to distribute and apply the ink also by cylinders. Another plan, more closely approaching that of the printing machines afterwards perfected by Mr. Napier and others, was to place the types upon a table and the paper upon an impressing cylinder, and to move the table backwards and forwards under it. In 1813, Messrs. Donkin and Bacon proposed placing the types upon a prism, which was to revolve against an irregularly shaped cylinder, on which the paper was to be placed. Nothing, however, could be effectually done in producing a proper working printing machine until the invention of inking rollers.

In 1814, Messrs. Bauer and Kœnig succeeded in constructing a machine, which was erected at the *Times* office, that produced 1,800 impressions an hour; and it continued in use till 1827. This rapidity of action, compared with that of the most improved printing press, produced a revolution in the art of printing; attention was then directed almost exclusively to the further improvement of the machines, and the platten press was neglected.

In the form of printing machines generally used, the types are laid upon an iron table that is moved to and fro by the turning of a wheel connected with a steam engine. The paper is placed upon cylinders covered with flannel, and the impression of the types is produced by the cylinders being fixed so closely to them that, as the table passes backwards and forwards, there is great pressure. The types are inked by a series of rollers, by which the ink is distributed and evenly laid on the face of the types without any manual labour.

The mechanical power gained by an arrangement of this kind arises from the pressure being exerted on a small surface at a time; consequently the power required for producing the impression of the types is not nearly so great as when the whole surface of the types makes the impression at the same instant. The force actually pressing on the types, from contact with the cylinders, is very much less than that brought to bear on them by the platten of the Stanhope press; but as it acts on a smaller surface at a time, the amount of pressure on each part, successively, greatly exceeds that received by any similar portion when it is impressed all at once. The difference of the action of the platten and of the cylinder may be compared to the different effects produced by a knife when pressed with its edge and with its flat side against a yielding surface ; the pressure on the flat surface may not be sufficient to leave any impression, whilst a much smaller pressure on the edge will produce an indentation.

The accompanying woodcut is a representation of one of Messrs. Applegath and Cowper's machines for printing both sides of the paper at the same time.

It consists of a cast-iron frame, about 14 feet long and 4 feet wide, on which the iron table, with the types upon it, slides backward and forward under two large cast-iron cylinders, covered with blankets, whereon the paper is laid. The pages of type to be printed on one side of the paper, and those pages that

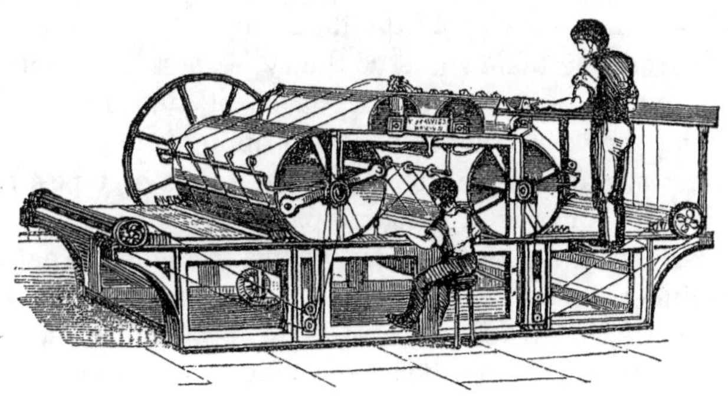

are to be printed on the back, are wedged into iron frames, called " chases," and these chases are fixed on the table at such a distance from each other, that they will pass under the two cylinders in the same relative positions. The sheets of paper are held on to the cylinders at their edges by means of tapes, and are so laid on by the workmen, that the type may be impressed on them with an equal margin all round. At each end of the machine is a supply of ink, which is taken from long iron rollers, about three inches in diameter, each of which turns in contact with a flat iron bar, that only allows a small quantity of ink to pass. A composition inking roller is made to vibrate between the inking table, where on the ink is thinly and evenly spread, and the iron feeding roller, and thus delivers the requisite quantity of ink on to the table. Several other composition rollers are placed across the inking table, with their axes resting in notched bearings, so that as the inking table moves forward and backward, those rollers distribute the ink evenly over it. There are four other rollers (none of which are shown in the diagram), which take the ink from the table; and as the types pass from under the cylinders, after printing a sheet, and return to them, they pass twice under the inking rollers. Each sheet of paper is laid by a boy on a web of tapes, by which it is carried round one paper cylinder, and then over and under two wooden drums to the other paper cylinder. The sheet of paper, in the course of its progress, is turned over, so as to receive the second impression on the other side; and as the tapes that carry it along leave the second cylinder, they divide, and the printed sheet falls into the hands of a boy.

In the printing machine which was shown at work in the Great Exhibition, invented by Mr. Applegath and made by Mr. Middleton, for printing the *Times*, the arrangements differ materially from those of the horizontal machines already described. The types, instead of being placed on a table, and moved to and fro under the impressing cylinders, are fixed to a large vertical cylinder, upwards of 16 feet in diameter, and there are eight impressing cylinders ranged vertically round it, with their axes fixed. By this arrangement there is no loss of time in withdrawing the types from under the cylinder to be again inked, but they move round from one fixed cylinder to another, receiving their ink between each, and thus producing eight impressions in succession during one complete revolution. At the *Times* printing office there are now three machines of that construction, two of which, with eight cylinders, print ten thousand an hour, and the other one, which has nine impressing cylinders, thirteen thousand.

The operations for printing that newspaper exhibit marvellous efforts of human ingenuity and skill, brought to bear in producing with the requisite rapidity a sufficient number of impressions to supply its enormous circulation. After the types have been composed and corrected, and ranged into columns and screwed up into their chases by upwards of one hundred hands, each page of type is attached to the large vertical cylinder—a curved form having been given to the type to adapt it to the circular surface. Nine men, standing each one beside a heap of damped paper, feed the largest machine by separating the sheets singly from the heap, and present them successively to the action of small rollers, that give each sheet a forward impulse, which brings it within the grasp of a series of endless tapes. These tapes catch hold of the sheets of paper, and carry them down to the level of the types. They are then shot along horizontally to the pressing rollers, covered with blankets, round which they are carried and pressed against the types; after which the endless tapes carry them away, and deliver them printed to a man below, who spreads them one over the other. A large reservoir of ink at the top of the machine supplies the inking tables, from which it is spread evenly over the inking rollers, and, at each revolution of the type cylinder, nine sheets are printed on one side. They are then taken to a second machine to be printed on the back, or, as it is called, "perfected." The accompanying engraving shows the general arrangement of the machines.

Few mechanical contrivances present so striking an illustration of the application of human ingenuity to the production of important results, and to the saving of labour, as these printing machines. To see the sheets of paper travelling along the tapes—to see them shoot downwards, carried sideways in one direction and back again, and delivered with half a million of words impressed upon them in less than three seconds, seems like the work of magic. To copy that number of words, thus printed in three seconds, would occupy a rapid penman forty days, working ten hours a day.

Great as are the printing powers of these machines of Mr. Applegath's, they have been surpassed more recently by one placed close beside them, invented by Mr. Hoe, of New York. In that machine the type cylinder is placed horizontally, by which means the paper is supplied directly to it without

altering its direction. As many as twenty thousand impressions in an hour have been produced by the

American machine, but it is not yet sufficiently perfected to be brought into regular use.

In another part of the *Times* establishment there is an ingenious machine for wetting the paper, by which contrivance much labour and time are saved. The paper, heaped in a pile at one end of a table, is presented in quires at a time to the action of a roller, which drags it on to a moving endless blanket, that is kept wet by a stream of water, and the upper surface is wetted by a long brush, placed over the blanket. The wetted paper is heaped upon a truck, which gradually descends, to keep the upper sheets on a level with the table, till the paper is piled up a yard in thickness. The truck is then raised, by hydraulic pressure, to the level of the floor, and is wheeled away and another one is loaded. Between nine and ten tons of paper are thus wetted daily; and the sheets of the *Times* printed during a year, if spread out and piled one upon another, would form a column as high as Mont Blanc. The quantity of ink daily consumed in the printing is upwards of two hundredweight. The machine is worked by two steam-engines, each of 16-horse power; and the noise of the numerous wheels and rapidly revolving cylinders is almost deafening.

The great rapidity and the comparative cheapness of printing by machines, together with the greater facility of making paper by machinery, have been the means of creating a demand for books which it would be impossible to supply, unless those means were at command. Thousands and hundreds of thousands of copies of publications, that spread knowledge among the people of the highest interest to the welfare of man, and replete with useful information of every kind, are now sold at prices which would be impossible, were it not for the improvements that have been made in the manufacture of paper, and in the means of printing.

Nor should we omit to notice, as one of the causes that have contributed to the production of cheap literature, the art of stereotyping, which has been perfected during the present century. Earl Stanhope, the inventor of the admirable press that bears his name, was prominent in bringing that art to perfection.

Numerous attempts had been made in the last century to produce casts from pages of type. So early, indeed, as 1700, some almanacks and pamphlets were printed in Paris from castings; and an edition of Sallust was printed in Edinburgh in 1739, from stereotype plates produced by Mr. Ged, a goldsmith. The process, however, was not encouraged, and on his death it was not further proceeded with. The most important advance in the art was made by M. Hoffman, of Alsace, who, in 1784, succeeded in obtaining stereotype plates by casting them in moulds of clay mixed with gelatine in which the pages of type were impressed, with which he printed a work in three volumes; but the castings were imperfect, and the plan was soon afterwards abandoned. Among the many plans tried to obtain perfect casts of the types when set up, was one contrived by M. Carez, a printer of Toul, who, in 1791, endeavoured to obtain casts in lead from a page of type, by allowing it to drop on the fused metal when it was in a state of setting. The matrices thus obtained were in like manner impressed on a fusible metal, which melted at a lower temperature than the lead. Good casts were often thus procured, but the uncertainty of the process, arising from the frequent fusion of the lead matrices, caused it to be discontinued. Other plans were tried in France with more or less success, but nothing was done practically until Lord Stanhope directed his attention to the subject in 1800, and resorted to the original method of obtaining matrices, by impressing the pages of type in a cold plastic substance. He employed plaster of Paris for his mould; and when they were thoroughly dried, they were plunged in fused type-metal; and in this manner a perfect cast in metal of the original page of movable type was produced. The process has been still further perfected, and casts from movable types, and from wood engravings, are now made with great facility, and the impressions from them are quite equal to the originals.

When it is intended to stereotype a work, the movable types used in composing it are new, and the "spaces" that separate the words from each other are longer than is customary when the type is to be printed from. These elongated spaces reach nearly to the face of the letters, so that the plaster may not sink between them. By this means the mould is easily removed from the face of the page of type. The metal casting of each page is very thin, and when required to be used, it is screwed on to blocks of wood to the same height as ordinary types.

Several attempts have been made to apply other substances than plaster of Paris and type-metal for stereotyping. At the Great Exhibition there were specimens of gutta percha stereotypes, that produced excellent impressions, and there were also fine stereotype castings of type in iron, from which a copy of the Bible had been printed. Papier maché has been found to be a material peculiarly applicable for the purpose, and it is now superseding the use of plaster of Paris for taking casts of the types.

By the application of the art of stereotyping, casts in metal of valuable works can be kept ready at any time, to be printed from when more copies are required; and the expense is saved of keeping on hand large stocks of printed paper, or of having a work re-composed when a further edition is wanted.

The inventions of Printing Machines and stereotyping were strongly opposed at first by pressmen and compositors, as calculated to diminish the demand for their labour. In "Johnson's Typographia," published in 1824, the "new-fangled articles" are mentioned in a spirit of great bitterness; and the writer thus poured forth his lamentations at the prospective ruin of the members of his profession:—"We are much surprised at the apathy and supineness shown by the body of master printers with respect to the subject under discussion; they most assuredly had good and sufficient grounds for an application to Parliament for a tax, that should bring the work so executed upon an equality with that done by manual labour."—"We feel satisfied that the above would not have met with encouragement from a British public, had they been aware of the evils attendant on it; they have not only to pay a full price for the work, but also extra poor's rates, in consequence of the men being thus out of employ; likewise they are countenancing the breaking up and destruction of all the energy and talent of that art which was England's proudest boast, and her shield against all the threats of her foreign foes."

These predictions of ruin have been completely falsified. It has been with the Printing Machines as with most other improved machinery for the saving of labour: on their first introduction some hands, no doubt, were thrown out of employ, but the advantages derived from the saving of labour very soon re-acted favourably in creating a greater demand for labour than before. The number of cheap periodicals, and the extensive issues of cheap literature in every form, require a much larger number of workmen to supply the demand, even with the aid of machinery, than was needed in the best days of the manual printing press; and at no time were so many compositors and pressmen employed as at present.

In the Reports of the Juries of the Great Exhibition, some interesting statistics are given, showing the influence of the invention of Printing Machines in extending the demand for books and periodicals. "The machine," it is observed, "created a demand,

and called into existence books which, but for it, would scarcely have been thought of. As the machine-work from type and woodcuts was far better than the ordinary printing of the day, booksellers were induced to print extensive editions, because they saw the machine could accomplish all they required. One of the first booksellers who availed himself of this power was Mr. Charles Knight, who projected the 'Penny Magazine,' on a hint from Mr. M. D. Hill, Queen's Counsel. Each number, published weekly, consisted of eight pages of letterpress, illustrated with good wood engravings. The public was astonished at the cheapness and good quality of the work, but it was its immense sale which rendered it profitable; for some years it amounted to 180,000 copies weekly. Mr. Knight, whose services in the cause of educational literature entitle him to the highest praise, expended £5,000 a year in woodcuts for this work. The Cowper machine has been the cause of the many pictorial illustrations which characterize so large a portion of modern publications. The 'Saturday Magazine,' 'Chambers' Journal,' the 'Magasin Pittoresque,' in France, and numerous others, owe their existence to this printing machine. The principle of *cheap editions and large sales* soon extended to established works of a higher value. A remarkable instance of this was the edition of Sir Walter Scott's Works, with notes, edited by himself; instead of the old price 10s. 6d., they were sold at 5s. a volume,* and the demand created by this reduction

* This statement does not adequately represent the reduction in

in price was so great, that, though the printer had a strong prejudice against machines, he was compelled to have them, the presses of his large establishment proving totally unable to perform the work, which amounted to upwards of 1,000 volumes per day for about two years. The Universities of Cambridge and Oxford have adopted Mr. Cowper's machines for printing vast numbers of Bibles, prayer-books, &c., &c. A Bible which formerly cost 3s. may now be had for 1s. Mr. Cowper recommended the Religious Tract Society to put aside their coarse woodcuts, to have superior wood engravings, and to print with his machine. The Society adopted those suggestions, and the result is, that by sending forth well-printed books, it could now support itself by their sale, without any aid from subscriptions."

As an illustration of the facilities afforded by the invention of Printing Machines in producing cheap editions of the writings of popular authors, the following curious facts relating to the Works of Sir Walter Scott, in addition to those furnished in the Reports of the Juries, may be found interesting.

In 1842, a general issue of these Works, in weekly sheets or numbers, at twopence each, was commenced by the late Mr. Robert Cadell, of Edinburgh, and completed in 1847. Of this edition, up to the present period (1858), the astonishing number of TWELVE MILLIONS OF SHEETS have been issued, the weight of

price; for each volume, sold at 5s., contained a volume and a half as originally published, besides Sir Walter Scott's notes; and the cheap volumes were illustrated with steel engravings.

which amounts to upwards of 335 tons! Another edition was published simultaneously by Mr. Cadell in monthly volumes at 4s., each containing about 360 pages; this series has reached a sale of more than 500,000 volumes. A third cheap issue, at eighteenpence a novel, is now being published by the present proprietors, Messrs. Adam and Charles Black, of Edinburgh. Nearly 300,000 volumes have already been printed of this edition.

It may be mentioned here, although hardly coming within the scope of the present article, but as affording a striking example of what literature has contributed to the revenue of the country in the person of a single author, that upwards of 3,500 tons' weight of paper * have been consumed in producing the various editions of Sir Walter Scott's Writings and Life; and the duty paid to Government on the paper, even at the present reduced rate, amounts to no less a sum than £51,450!

Since the Juries made their Reports, the development of cheap literature has been greatly extended. Newspapers, some of which contain eight full-sized pages, of six columns each, printed in small type, are sold for the marvellously low price of a penny, and are stated to issue as many as 50,000 copies daily; and some of the newspapers and other periodicals, printed on good paper, are issued for a halfpenny. Among the works of a standard character, published at prices which nothing but a very extensive scale

* If the number of sheets of paper used in printing these works were laid side by side, they would extend nearly *fifty thousand miles!*

could make remunerative, may be mentioned the popular series which includes "The Reason Why," and "Enquire Within upon Everything." Of the eight volumes already issued, each containing about 350 closely printed pages for half-a-crown, nearly 170,000 copies have been sold within a period of less than three years.

LITHOGRAPHY.

The art of printing from stone was invented at the end of the last century by M. Aloys Senefelder, of Munich; but it was not brought to such a state of perfection as to be practically useful until many years afterwards.

The principle on which Lithography depends is the different chemical affinities of water for oily and for earthy substances, which cause it to run off from the one and adhere to the other. The drawing or writing is made in oily ink upon a smooth calcareous stone that will absorb water, so that, when the stone is moistened, the water adheres to it and leaves the lines of the drawing traced upon it dry. An inking roller, charged with an oily ink, is then passed over the stone and inks the drawing, but leaves all the other parts of the stone quite clean. A damped paper is next laid on, and when subjected to great pressure, an exact copy of the drawing or writing is produced.

This simple and ingenious process has been carried to such perfection, that the most beautiful artistic effects can be produced by it far more economically than by any other style of engraving; and further improvements in the art are being continually made. It is satisfactory, therefore, to be able to trace its history from its very beginnings, of which an interesting account has been published by the inventor himself.

M. Senefelder's father was an actor at Munich, and in his youth he followed the same profession. He turned his attention afterwards to music; and it was in his attempts to devise some means of printing his compositions economically that he chanced to discover the art of Lithography.

He had previously made himself acquainted with the methods of copper-plate printing, and he commenced his operations by etching the notes of music on copper-plates, covered with varnish in the ordinary way. He found, however, that it would require much practice to enable him to do this properly, and not being able to buy copper-plates for his rude essays, he thought of practising upon stones. Fortunately for the success of his efforts, the quarries at Solenhofen, near Munich, supplied him with slabs of stone admirably adapted for the purpose; and it is a remarkable coincidence, that the material which Senefelder used for his experiments is the best for the purpose of Lithography that has hitherto been discovered. His chief object in making use of these slabs of stone was to practise himself in the manipulation of writing the notes, and of biting them in with *aqua-fortis* (nitric acid), as he supposed the slabs would be too brittle to bear the action of the press. He did not try, therefore, to have these etchings on stone proved by the press, but he contented himself with holding them up to a mirror to observe the progress he was making in writing backwards.

Having at length been supplied with much thicker slabs of stone, to bear the requisite pressure, he endeavoured to grind and polish them sufficiently for the purpose of being printed from, in the same manner as copper-plates. He succeeded to some extent in doing so, by means of diluted nitric acid; and he contrived to obtain about fifty good impressions from the stone.

In all these attempts at Lithography, the lines were etched into the stone by the action of nitric acid, and the only advantages professed to be gained by the process were the questionable ones of comparative cheapness of material, and greater facility of working. M. Senefelder admits that there was nothing new in engraving upon stone; all that he claims in that part of the invention is, the manner of polishing the surface, and the composition of the ink adapted for printing from it. The most important step in the progress of the invention of Lithography, as at present practised, was made by accident, which he thus describes:—

"I was preparing a slab of stone for engraving, when my mother asked me to write a memorandum of things she was about to send to be washed. The washerwoman was waiting impatiently whilst we searched in vain for a piece of paper, and the common writing ink was dried up. Having no other writing materials, I wrote the washing bill on the stone I was about to prepare for engraving, using for the purpose my ink made of wax, soap, and lamp-black, intending to copy it afterwards on paper. Whilst looking at the letters I had written, the idea all at once occurred to me how it would do to cover the stone, with the writing upon it, with aqua-fortis, so as to leave them in relief, and then to print from them in the same manner as woodcuts, with a common letter press. The attempts I had hitherto made to engrave upon stone had taught me that the relief of the letters thus obtained would not be much. Nevertheless, I made the attempt. I mixed one part of aqua-fortis with five parts of water, and poured it on the stone to the height of two inches, having previously walled it round with wax in the usual manner. The diluted aqua-fortis was permitted to rest on the stone five minutes. I then examined the effect, and I found that the letters were raised above the stone about the thickness of a card. Most of the lines were uninjured, and retained their original size aad thickness. This gave me the assurance that writing, sufficiently traced, especially if the letters were in printed characters, would have still greater relief."*

Though M. Senefelder had advanced thus far, he had not yet made application of the chemical properties of ink and water, which constitute the distinguishing characteristics of Lithography. That was

* "L'Art de la Lithographie;" par M. Aloys Senefelder, Inventeur de l'Art Lithographique. Munich, 1859.

reserved for a further discovery, also brought about by accident. The difficulty he experienced in writing words on the stone in the reverse way, induced him to adopt the plan of writing the letters on paper with a soft black-lead pencil, and then transferring them on to the stone by pressure. He subsequently used lithographic ink for the purpose; and in the course of his experiments he observed, that when a paper written on with lithographic ink, and well dried, was dipped into water on which some oil was floating, the oil adhered to the writing, and left the rest of the paper clean, and that this effect was most striking when the water contained some gum in solution. This discovery induced him to try the effect on printed paper; and, taking a printed page from an old book, he moistened it with gum-water, and afterwards sponged the whole surface with oil colour. The colour adhered to the letters, and left the paper clean, and after further experiments he succeeded in printing as many as fifty copies from a page of printed paper; the letters, of course, being reversed. The idea then suggested itself of transferring, on to stone, letters written with lithographic ink upon paper. The plan succeeded, and the principle of the art of Lithography was thus applied to practice. M. Senefelder, in his subsequent improvements, gave a slight relief to the letters by the original plan of using diluted aqua-fortis, by which means the impressions obtained were blacker. He also contrived the means of printing in colours from stone, by reversing the process of ordinary lithographic printing. To produce coloured prints, he left uncovered all the parts that were to receive the colour, and the other parts of the stone were covered with an oleaginous fluid, that quickly dried. On applying any water-colour to the stone, it adhered to the uncovered surface, and not to the covered parts, and that colour was transferred to paper by pressure. In this manner, by using several stones properly prepared, the different colours required were printed, and the effect of a coloured drawing was produced. Thus we perceive, that almost at the first invention of the art of Lithography, the ingenious inventor showed the way of applying it to the production of coloured prints, a process which has lately been carried to great perfection.

Senefelder lived to see his invention extensively adopted, and to reap benefit from his ingenuity. He died at Munich, in 1834, after having been many years the director of the Government lithographic office; and, in the latter years of his life he received a handsome pension from the King of Bavaria.

There is little to be added to the description of the process of Lithography, beyond that given by the original inventor in 1819, the principal advances that have been made in the art having consisted in improved methods of manipulating. The ink now generally employed for drawing on the stone consists of equal parts of tallow, wax, shell-lack, and soap, mixed with about one-twentieth part of lamp-black; but the composition is varied, according to the kind of design to be executed. For writing or drawing upon paper, to be transferred to the stone, more wax is added to the ink, to give it greater tenacity.

The drawing upon paper, to be transferred to stone, is not attended with any difficulty, and may be done by ordinary artists. The ink is applied with a pen, or camel's hair pencil, and when the effect of chalk drawings is required to be imitated, the ink is shaded by means of stumps, similar to those used in chalk drawings on paper. Some artists prefer to work directly on the stone with a camel's hair pencil, or with a composition called lithographic chalk.

To transfer the drawing from paper on to the stone, the paper is first sponged with diluted nitric acid, which decomposes the size, and renders it bibulous. After being placed for an instant between blotting paper, to remove superfluous moisture, it is laid with the drawing downwards on the stone, which is slightly warmed. The stone is then passed through the press, and the drawing adheres firmly to it. To remove the paper, it is wetted at the back with water, and, when quite soft, it is rubbed with the hand. In this manner every particle of the fibrous pulp is cleared away, and the drawing or writing in ink remains as if it had been drawn directly on the stone. To prepare the stone for taking the ink, gum water is poured upon it, and it is rubbed over with a rag containing printer's ink, which serves to blacken the writing and prepares the lines for afterwards receiving the ink.

The lithograph thus prepared is given to the printer, who first etches it, in the manner originally practised by M. Senefelder. The nitric acid employed for the purpose is diluted with about thirty parts of water, and it is poured over the stone whilst it is inclined on one side. This process is repeated several times, the object of it being not so much to give relief to the lines, as to roughen the surface of the stone, and thus facilitate its absorption of water. The nitric acid also removes the alkali from the drawing ink. In printing, gum is added to the water with which the stone is moistened, as an additional preventive of the ink adhering to those parts not drawn upon. The printing ink is applied with large rollers, and the damped paper having been placed carefully upon the stone, with blankets at the back, it is passed through the press.

The lithographic press somewhat resembles in form an iron printing press, but differs from it greatly in its mode of action. Instead of the large flat plate that in a printing press is pulled down upon the whole surface of the types, a long, narrow arm, called a scraper, is brought to bear upon the stone, and the table whereon the stone is laid is pushed forcibly under it, by which means a great pressure is exerted on a smaller surface at successive times, instead of being brought to bear all at once. In the principle of its action, indeed, a lithographic press is like a printing machine, and steam lithographic presses have been invented to work in a similar manner, though the practical results have not generally been very successful.

Among the many applications of lithography, the transfer of copper-plate engravings is one of the most useful. An impression of the plate is taken on paper that is coated with a compound of flour, plaster of Paris, and glue, and from the paper it is transferred to stone. By this plan the original plate remains untouched, and the printing from the stone is much cheaper than from the copper. Tinted lithography and chromo-lithography, by which the beautiful effects of coloured drawings are produced in the manner indicated by M. Senefelder, have recently been applied very successfully.

AERATED WATERS.

THE invention of soda-water, in the state in which it is now known, as an effervescing beverage impregnated with three or four times its volume of carbonic acid gas, is of very modern date. There are, indeed, to be found in most of the old works on chemistry descriptions of Nooth's apparatus for impregnating liquids with carbonic acid; but all that was attempted to be done by that apparatus was to produce an impregnation of the water with little more than the quantity of gas it will naturally absorb under the pressure of the atmosphere. It was not until about the year 1815 that mechanical pressure was applied to force a larger quantity of gas into combination with water, to imitate the briskly effervescing medicinal waters of Germany.

Mr. Schweppe and Mr. Paul were the first who introduced the manufacture of artificial effervescing waters into England, and soda-water was then considered, as tea was on its first introduction, entirely medicinal. Indeed, the quantity of soda which was at that time usually dissolved in the water gave it a disagreeable taste; but when the manufacturers diminished the quanity of alkali, and increased the volume of gas forced into the water, they produced a pleasant beverage, which soon became in request for its refreshing, wholesome qualities.

The apparatus for the manufacture of soda-water, as it is usually made on a large scale, consists of a strong vessel, furnished with a safety valve, in which the water is impregnated with gas. This vessel, containing about nine gallons, is made of thick wood, well seasoned and nicely fitted, and bound round with strong iron hoops, the heads of the cask being well secured by means of iron bolts and screw nuts. It is requisite that the receiver should be capable of bearing a pressure of at least six atmospheres, which is equal to 90 lbs. to the square inch.

The carbonic acid gas is generated from chalk or whiting and diluted sulphuric acid. The materials are mixed together in a small closed wooden or leaden vessel, provided with an agitator, that can be worked by a handle fixed to a projecting axis at the top. The gas, as generated, enters by a bent tube into a gas-holder, the opening of the tube being under water. By this means the gas is freed from the fumes of sulphuric acid vapour, and from the fine particles of chalk that become mingled with it during its sudden liberation. The gas sometimes undergoes a further purification, by passing through a gas washer, before it is forced into the water.

A small force-pump, worked by a crank, with the assistance of a fly-wheel, draws the carbonic acid from the gas-holder, and forces it into the water. The combination of the gas and water is facilitated by an agitator, the axle of which projects through a stuffing box, and it is worked either by hand, or is turned by means of a small cog-wheel, that works into the teeth of a larger one fixed to the crank axle, so as to produce rapid rotation.

It is found requisite, in the first place, to expel the atmospheric air in the receiver; for which purpose the safety valve is left open for a short time after the gas is being forced in, otherwise it would retard the impregnation of the water by the gas. When the gas and water are well incorporated, the liquid will contain as many volumes of gas as there are atmospheres of pressure in the air-space above it in a state ready to effervesce, and one other volume, with water absorbs under the pressure of the atmosphere. Thus, when there are three atmospheres of gas under pressure, each bottle of soda water contains four bottles full of gas, which are absorbed without perceptibly increasing its bulk. The perfect impregnation of the water with gas, however, requires time. The water will, indeed, become brisk almost as soon as two or three atmospheres of gas have been forced in, but it will not acquire the flavour of good soda-water until the gas and water have been allowed at least half an hour to digest; and it is improved by remaining in contact for several hours.

The temperature has considerable influence in the process of impregnation, for in hot weather the gas will not combine so readily, nor will the water absorb an equal volume of gas. In summer time, therefore, soda-water should be made before the heat of the day, and ice should be added to the water.

When the receiver is fully charged, and the operation of bottling begins, every bottle-full that is drawn off diminishes the pressure on the water that remains; and if no means were taken to add more gas, the soda-water would gradually become weaker and weaker as each bottle was drawn off. It is usual, in the best arranged apparatus, to have two tubes connected with the force-pump, one of which feeds it with water, the other with gas, by which contrivance water and gas, in their proper proportions, are continually forced into the receiver, which may thus be always kept nearly full.

The process of bottling requires great manual dexterity. The neck of the bottle is pressed by a lever against a collar of leather fixed to a flange on the tap, so that, when the soda-water rushes in, no air nor gas can escape. The pressure inside the bottle therefore quickly becomes equal to that of the receiver, and the water ceases to flow through the tap, until some of the air is allowed to escape. When the bottle is nearly full, the operator quickly withdraws it with one hand, and having a cork ready in the other, he puts it in before the water can rush out. The cork is then forced in further by pressure, and fastened down by wires or strings.

A bottling apparatus has been invented for facilitating the process; but a man accustomed to bottle by hand can do it more quickly, and with as little waste of gas and water as with a machine. Much depends, however, upon the state of the soda-water in the receiver; for if the gas be well digested, and the temperature low, it rushes into the bottle with much less force, though the water may contain a greater quantity of gas, than when it is newly made, and apparently more brisk. The bottles very frequently burst during the operation with great violence, and unless they are enclosed in a guard, the men are liable to be severely injured. Glass bottles have now generally supplanted those made of earthenware, which were formerly used; and though the glass bottles are much stronger than the earthenware ones, the bursting of them, when it does occur, is far more dangerous.

The process of forcing gas into the water by mechanical pressure, in the manner described, requires great labour, for the pump has to be worked against a pressure exceeding fifty pounds on the square inch. With a view to remove that inconvenience, and to avoid the use of costly machinery, so that private individuals might manufacture soda-water, the author contrived a modification of Nooth's apparatus, for which he obtained a patent in 1831. By that means, the gas is generated in a closed vessel, and forces itself into the water by its own elasticity. Any amount of pressure can thus be obtained by chemical action alone. The accompanying woodcut represents a section of the apparatus in its improved form. The vessel, A, is made of very strong stone ware, inside which is the gas generator *b*. A few inches from the bottom of the generator is the partition, *a*, perforated with holes, and near the top there is inserted the small tube, *c*, which conveys the gas down to a perforated expansion of the tube, *d*, through the apertures of which the gas issues into the water contained in A. Another tube, *e*, reaches near the bottom, and is connected

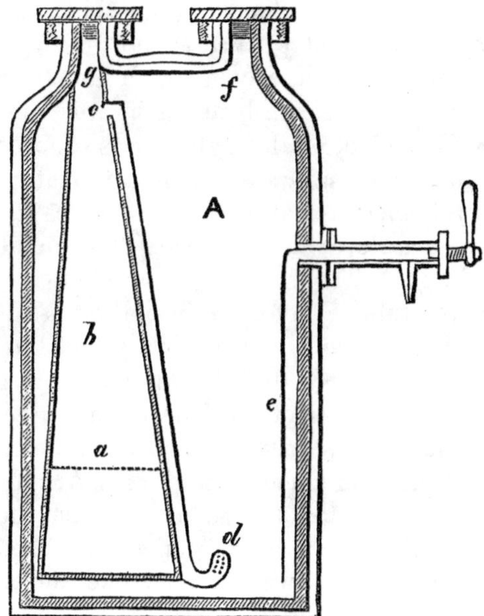

with a stop-cock for the purpose of drawing off the aerated liquid. In charging the apparatus, the interior, A, is nearly filled with water, or other liquid, through the opening, *f*, which is then closed by cork, which is kept in its place by a screw nut. A few ounces of carbonate of soda, mixed with water, are then poured into the generator through the opening at *g*. The mixture flows through the apertures in the partition, and occupies the lower part of the generator. A proportionate quantity (about three-fourths of the weight of the soda) of tartaric acid in crystals is then introduced through *g*, which lodge on the top of the partition without touching the soda. The opening being then closed by a screw-nut, the apparatus, which is mounted on pivots, with an appropriate stand, is swung backwards and forwards like a pendulum. The effect of this agitation is to force a portion of the water saturated with carbonate of soda through the apertures at *a*, where it comes in contact with the tartaric acid, and instantly generates carbonic acid gas. The gas, having no other escape than through the tube, *c*, is forced into the vessel A, and becomes mingled with the water by the same act of vibration that brings the soda and tartaric acid together. The continuance of the vibratory action for a short time generates sufficient gas to aerate the water or other liquid contained in the vessel, A. When the aeration is completed, the soda-water may be drawn off, as required, through the stop-cock. The apparatus is made of two sizes, to hold one and two gallons.

The tartaric acid and soda in the generator do not mingle with the water, and the tartrate of soda, resulting from the combination, is emptied after the soda-water is drawn off, before renewing the charge.

A French modification of this apparatus, in glass

vessels protected by cane netting, called a "gasogene," has recently been introduced, and is extensively used. The materials for generating the carbonic acid gas are put into the smaller vessels, and kept separate until the apparatus is inverted, and then gas is rapidly generated, and forces itself through the water.

The powders that are sold for making soda-water, by mixing them together, consist of carbonate of soda and tartaric acid. When brought together in solution, a violent effervescence ensues, but the gas is not combined with the water in the same manner as when it is forced in and allowed to remain for some time with the liquid to be aerated. There is the further disadvantage attending such powders, that the tartrate of soda, formed by the tartaric acid and the carbonate of soda, employed to generate the gas, is drunk with the water.

REVOLVERS AND MINIE RIFLES.

"Is there anything whereof it may be said, See, this is new? it hath been already of old time, which was before us."* This observation of Solomon, the correctness of which we have often seen verified in this History of Inventions, is applicable even to that great apparent novelty the formidable "Revolver"—that death-dealing weapon, which will fire six shots in rapid succession by merely pulling the trigger so many times, as fast as it is possible.

The Revolver was almost unknown in this country until 1851, when it was brought prominently into notice at the Great Exhibition, by the specimens shown there by Colonel Colt, of the United States. Pistols with six barrels, which might be fired successively with the same lock, by turning them round, were, indeed, previously seen in gun-shops; but their clumsy form and their great weight prevented them from being used. Nor was Colonel Colt much more successful in his earlier attempts to bring his Revolver into public notice. He obtained his first patent in

* Book of Ecclesiastes i. 10.

America in 1835, and established a manufactory for the pistols at Paterson, United States, where he expended £35,000 in attempting to bring the fire-arm to perfection, but with no beneficial result to himself beyond gaining costly experience. He made further improvements in 1849, and so far perfected the weapon that it had been used extensively in America before it was brought into notice in this country.

When Colonel Colt came to England, he undertook to investigate the origin of repeating fire-arms, with a view to ascertain how far he had been anticipated; and the result of his researches was, that repeating fire-arms, similar in principle to his own Revolver, had been invented *four centuries before.*

He found in the Armoury of the Tower of London a matchlock gun, supposed to have been made as early as the fifteenth century, which very closely resembles, in the principle of its construction, the Revolver of the present day. It has a revolving breech with four chambers, mounted on an axis fixed parallel to the barrel, and on that axis it may be turned round, to bring any one of the four loaded chambers in succession in a line with the barrel, to be discharged through it. There are notches in a flange at the fore end of the revolving breech to receive the end of a spring, which is fixed to the stock of the gun, for the purpose of locking the breech when a chamber is brought round into the proper position. The hammer is split at the end, so as to clasp a match, and to carry its ignited end down to the priming powder when the trigger is pulled. Each chamber is provided with a priming pan that is covered by a swing lid, and, before firing, the lid is pushed aside by the finger, to expose the priming powder to the action of the lighted match. If the date of this gun be correctly stated, a very rapid advance in the art of gunnery must have been made after the invention of gunpowder, which took place only one hundred years previously. The want of a better mode of discharging the gun than a lighted match was one of the chief obstacles to the introduction of the Revolver four centuries ago.

There is also in the Tower Armoury a specimen of a repeating fire-arm of a more recent date, though still very ancient, that presents considerable improvement on the preceding one. It has six chambers in the rotating breech, and is furnished with a barytes lock and one priming pan, to fire all the chambers. The priming pan is fitted with a sliding cover, and a vertical wheel with a serrated edge projects into it, nearly in contact with the powder in the pan. To this wheel a rapid motion is given by means of a trigger-spring, acting upon a lever attached to the axis of the wheel; and the teeth of the wheel strike against the barytes, which is brought down, previously to firing, into contact with it, and the sparks thus emitted set fire to the powder in tbe priming pan, and discharge the piece. In this instance, also, the breech is rotated by hand.

A still further advance towards perfection in repeating fire-arms is to be seen in the United Service Museum, where there is a pistol, supposed to have been made in the time of Charles I., with the breech rotated by mechanical means. In this pistol, the act of pulling back the hammer turns the breech, containing six chambers, one-sixth part of a revolution, and the priming powder is ignited by a flint hammer striking against steel.

The manufacture of these fire-arms presented some practical difficulties which could only be overcome by great care and skill in the construction; and from the failure in this respect they were not patronized. It was necessary, in the first place, that the loaded chambers should be brought into an exact line with the barrel, and be firmly retained there during the discharge. It also required great nicety in the fitting of the breech to the barrel, to prevent the fire from communicating to the other chambers. A further difficulty was to prevent the spindle, whereon the breech revolved, from becoming foul by the explosion of the powder; otherwise, after firing a few times it would stick fast, and the gun would become useless.

The earliest patent for repeating fire-arms in this country was obtained by James Puckle, in 1718, for a gun with a rotating breech. There were six chambers in the breech, which was turned round by a winch, and, when the six were fired, there was an arrangement by which the chambered breech could be removed, and another loaded one substituted for it. Mr. Puckle appears to have been of a poetical turn of mind, and the specification of his patent is enlivened by the following loyal couplet, which deserves to be quoted as a novelty in patent records :—

> "Defending King George, our country and laws,
> Is defending yourselves and the Protestant cause."

The invention of percussion priming in 1800, by the Rev. A. J. Forsyth, was an important step towards the perfection of fire-arms generally, and of Revolvers in particular; for until the chambered breech could carry round with it in a compact form the priming for each chamber, the construction must have been clumsy, and the action uncertain.

Colonel Colt, as already stated, took out his first patent in 1835, and in 1849 he patented the improved Revolver, which he has brought into general use. It has six chambers in the rotating breech, and the nipples to hold the percussion caps are sunk into a recess, so that the lateral fire, if any, cannot reach them; and at the other end, the chambers are protected from lateral fire by chamfering their mouths. By these means, the danger of firing the gunpowder in the other chambers is effectually provided against.

The demand for Colt's Revolvers became so great after the last improvements were made, that at his manufactory, at Hartford, in America, he made 53,000 of them in 1853; and at his manufactory at Vauxhall, near London, he employs upwards of 300 workmen, though by far the largest portion of the work is done by machinery.

Several improvements have been introduced in Revolvers since Mr. Colt's patent of 1849, among which is the arrangement, made by Mr. Adams in 1851, for causing the chambered breech to turn by the action of pulling the trigger, which at the same time draws back the hammer. By this arrangement, the whole of the six loaded chambers may be discharged in three seconds, whilst the pistol continues presented.

The latest improvements in Revolvers were contrived by Mr. Josiah Ells, of Pittsburg, North America, as specified in a patent obtained for him by the author, in his own name, in 1855. The annexed woodcuts show the figure of this Revolver, with the

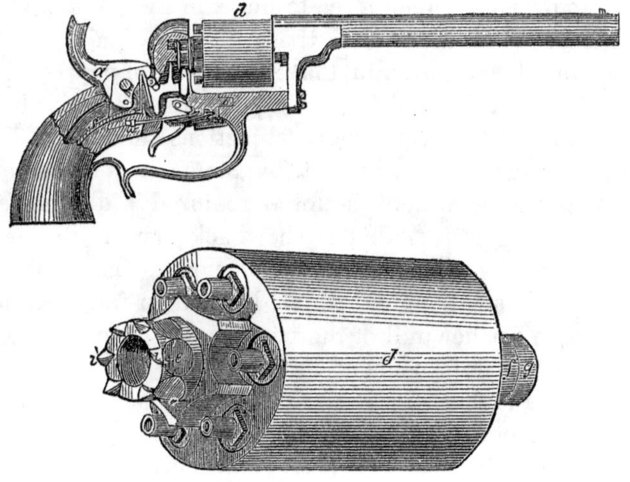

working parts round the lock exposed to view, together with the shape of the revolving chambered breech.

In this improved Revolver, the force required to pull back the hammer, *a*, is regulated by a double spring, *w*, so as to diminish as the hammer is drawn back; and at the moment of firing a slight pull of the trigger is sufficient. Another improvement consists in the addition to the chambered breech, *d*, of a projecting tube, which prevents the spindle on which it turns from becoming foul; and there is also a safety bolt added, as a protection against accidental firing.

The plan of making the mere action of drawing the trigger turn the chambered breech and pull back the hammer, as originally contrived by Mr. Adams, required so much force to pull the trigger as to interfere materially with the accuracy of aim. There was danger, also, in that mode of turning the chambered breech, arising from premature firing. In Mr. Ells's Revolver these objections are in a great measure obviated; first, by the action of the double spring, by which the force required is diminished as the trigger is pulled farther back; and in the second place, by making the shoulder of the hammer catch into a small notch, which holds it at full cock, until, by a further pull of the trigger, the pistol is fired.

An improvement in the art of war, no less important than the Revolver, was introduced nearly at the same time. The Revolver affords a formidable means for attack or defence at short distances, whilst the Minié Rifle extends the destructive range of firearms far beyond the distance to which the ordinary musket ball could reach. The principle of rifling gun barrels was first made known in the specification of an invention patented in 1789, by Mr. Wilkinson, the improvement he effected being thus described:—

"The gun, or piece of ordnance, after being bored in the usual method, hath cut therein two spiral grooves, which run the whole length of the bore. These curves, according to their curvature, will give a circular motion to the shot during its flight."

The spiral grooves, when the bullets are rammed down, cause the ball to offer greater resistance, therefore the explosive force of the gunpowder is brought to act upon them more completely before they leave the gun barrel; and the rotary motion imparts greater steadiness to the ball. Rifled barrels, therefore, carry the balls farther, and increase the accuracy of the aim. They, however, require increased power and longer time to ram down the ball in loading, and the risk of bursting the gun is increased if the ball be not rammed close upon the powder. For these reasons, they were considered unfit to be employed generally by soldiers, and they were entrusted only to select corps of rifle shooters. The object of Captain Minié's invention was to facilitate the loading of rifles, by contriving a bullet which might be easily rammed in, and would expand in the act of firing, so as to fill up the grooves. What is commonly called the Minié Rifle is, in fact, only a Minié Rifle Ball, for the barrels of the guns are nearly the same as the ordinary grooved rifles.

The ball is an elongated one, with a hollow cone at the bottom, into which is fixed an iron button. When the gun is fired, the button is forced into the cone, and expands the lead, which thus fills up the grooves and gives a spiral direction to the bullet.

The Minié ball serves the purpose excellently for a short time, but after firing several rounds the iron button is forced through the lead, leaving a portion of it behind, which clogs up the barrel, and renders it unfit for use.

Several substitutes for iron were tried, to remove that inconvenience, and it was at length found that the button might be dispensed with altogether, for the hollow cone is of itself sufficient to expand the lead. The balls are, therefore, now made in that manner at the Government gun manufactory at Enfield, and the rifled guns now used in the army, which carry bullets to the distance of a mile and more, are called the *Enfield Rifle*. The cost of each of these rifles to the Government is stated to be £3 4s. 7½d. As the balls are made to slip into the barrels easily, they can be loaded as readily as the common musket: and they will carry three times the distance, with much more certainty.

CENTRIFUGAL PUMPS.

Many ingenious men have vainly attempted to apply what has been erroneously called "centrifugal force" as a motive power, conceiving that the effort made by bodies to fly off when whirled round in a circle was occasioned by a force generated by their rotation. The experiment of the "whirling table," which is commonly shown to illustrate centrifugal action, tends to confirm the notion that force is generated; for it is there seen that, when the velocity of rotation is doubled, the centrifugal force is quadrupled, and that it continues to increase in a geometrical ratio. It has, therefore, been conceived that a power might be generated of indefinite amount; for as a double velocity can be communicated by doubling the moving power, whilst the tendency to fly off at the circumference is quadrupled, there appeared to be a creation of power which, if properly applied, would realize perpetual motion.

A working engineer known to the author was so fully possessed with the notion that power might thus be created, and that its application would be of the utmost benefit, that he imagined he had been specially appointed to reveal the principle to man, as a boon of inestimable value to the manufacturing arts. The plan he adopted was to employ what he called a generating engine, consisting of a centrifugal pump; and the force with which the water was projected from the ends of two rotating horizontal arms was directed against pistons working in cylinders, as the force of steam is in a steam engine. Having once set this machine in action, he expected to be able, by means of the self-creating centrifugal force, to generate the power that worked the generating engine, and thus to have a reservoir of force of any magnitude constantly at command. So completely satisfied was he of the practicability of the plan, founded, as he supposed, upon one of Newton's laws of motion, and he felt so happy in the thought of being charged with an important mission for the benefit of mankind, that it was almost cruel to attempt to correct his notions of the power of centrifugal force. He spent all his money in endeavouring to realize this impossible project, and even its failure did not convince him of his error.

The simple kind of Centrifugal Pump applied in that chimerical scheme was known upwards of one hundred years ago. It consisted of a vertical hollow shaft, into which were inserted two horizontal arms. The shaft was supported on a pivot at the bottom, and was turned by a handle at the top, as represented in the accompanying drawing. The lower end of the vertical shaft was immersed in water, and when rotary motion was given to the machine, the centrifugal action propelled the water from the ends of the arms, and the water rose in the vertical shaft to supply its place.

The effect in a pump of this construction is due to the pressure of the atmosphere, for the outpouring of the water from the rotating arms tends to produce a vacuum in the shaft, in the same manner as the lifting of the plunger in a common pump. It is

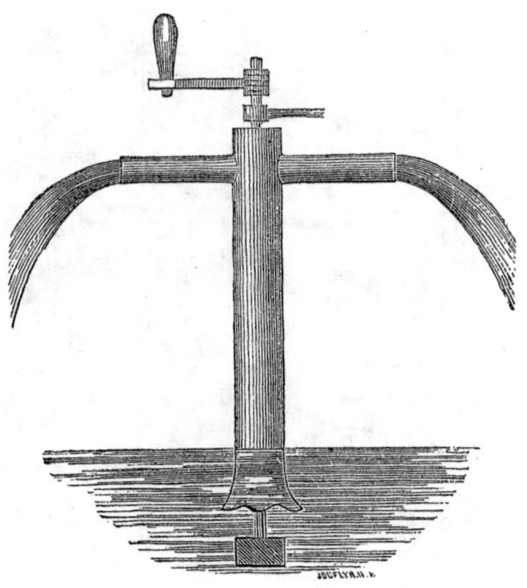

evident, therefore, that a Centrifugal Pump of that construction could not raise a column of water higher than the pressure of the atmosphere would force it up, which would be about thirty feet.

Mr. Appold's Centrifugal Pump, which constituted one of the most remarkable features of the Machinery Department of the Great Exhibition, is constructed on a different plan, though the principle is the same. The rotating arms are immersed in the water to be raised, and to diminish the resistance which would be produced by the rotation in water of

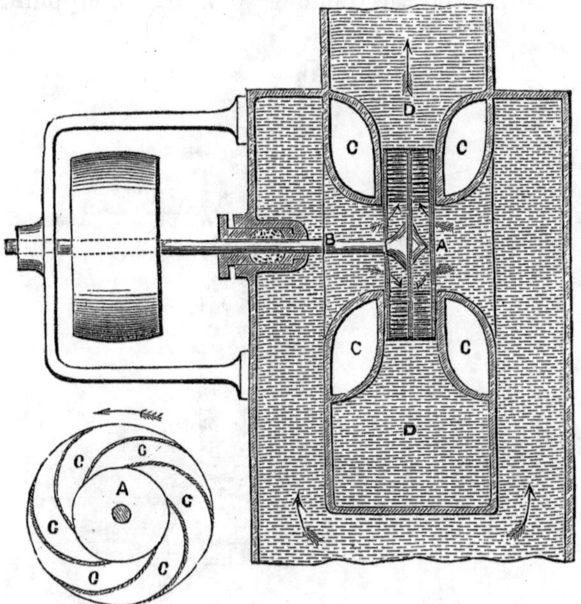

two or more exposed arms, they are enclosed within discs of metal, about one foot in diameter, and three or four inches apart. The arms are formed by curved partitions between the discs, which radiate from the centre to the outer rim, towards which the space between the discs is contracted. This pump is fixed on an axis, to which rapid rotary motion can be given; and it is fitted into a case connected with the pipe that conveys the water to the discharging orifice. The water enters the rotating disc through a large aperture in the centre, and it is forced through the spaces formed by the radical arms with increasing velocity, until it escapes from the circumference. Sections of Mr. Appold's pump are shown in the accompanying diagrams, in which A is the central opening for the admission of water; C, C, C, the curved radical partitions which form the arms by which motion is communicated to the water, and through the ends of which it issues into the external case, connected with the lift-pipe, D.

In the Great Exhibition there were two other Centrifugal Pumps shown in action, one by Mr. Bessemer, and one by Mr. Gwynne, from America; but neither of them exhibited such striking effects as Mr. Appold's, which was so arranged as to throw out a continuous cascade of water from an apertnre six feet wide, at a height of twenty-six feet. The Jury of Class V., who made numerous experiments to determine the practical efficiency of Centrifugal Pumps, and the relative merits of the three exhibited, reported very favourably of that of Mr. Appold, to whom a Council Medal was awarded. When rotating at the rate of 788 revolutions in a minute, and lifting the water 19·4 feet, the greatest practical effect, compared with the power employed, was attained. The discharge of water per minute at that height, with the pump rotating with a velocity of 788 revolutions, was 1,236 gallons; and with a lift of 8 feet, 2,100 gallons per minute were discharged, when the rotating velocity was 828 revolutions per minute. In Mr. Gwynne's and Mr. Bessemer's pumps, which had straight vanes, the ratio of power to the effect did not exceed 0·19. One of Mr. Appold's pumps, only one inch in diameter (the exact size of the small diagram), will discharge ten gallons per minute. The greatest height to which water has been raised by the pumps that are one foot in diameter is 67·7 feet, with a volocity of 4,153 feet per minute.

The velocity with which the pump should revolve depends upon the height to which the water is to be raised. Beyond a certain height, the required velocity is practically unattainable, but long before that limit is reached the waste of power becomes so great, that the pump is of no value, for the pressure on the circumference counteracts the force with which the water is expelled. It is, therefore, only at comparatively low levels that the Centrifugal Pump is a useful engine. The absence of all valves renders it very suitable for draining marshes, and for other similar purposes, as the muddy water and suspended matters will not obstruct its action.

In the Report of the Jury the influence of the curved shape of the radial arms is considered very important in producing the effects. "If the vanes be straight," the Report observes, "it is evident, that whatever may be the velocity of the water in the direction of a radius, when it leaves the wheel its velocity in the direction of a tangent will be that of the circumference of the wheel, so that the greater the velocity of the wheel the greater will be the amount of *vis viva* remaining in the water when discharged, and the greater the amount of power uselessly expended to create that *vis viva.* If, however, the vanes be curved backwards as regards the motion of the wheel, so as to have nearly the direction of a tangent to the circumference of the wheel at the points where they intersect it, the velocity due to the centrifugal force of the water carrying over the surface of the vane in the opposite direction to that in which the wheel is moving, and nearly in the direction of a tangent to the circumference, will—if this velocity of the water over the vane in the one direction be equal to that in which the vane is itself moving in the other—produce a state of absolute rest in the water, and entire exhaustion of *vis viva.*" It is an interesting fact in the history of the invention, that the curved form was formerly adopted in some of the American pumps, and afterwards abandoned.

There are competing claims to the invention of Centrifugal Pumps in the form now adopted. This kind of pump is stated to have been used in America in 1830. M. Charles Combe took out a patent in France for a similar pump in 1838; but though Mr. Appold was later in the field with his more perfect machine, he appears to have proceeded independently of previous inventors.

TUBULAR BRIDGES.

No sooner had the formation of railways commenced for carrying passengers in long trains of carriages drawn by heavy locomotive engines, than the want was experienced of some different kind of bridge from any then existing for crossing rivers, roads, and valleys. The train could not be turned sharply round a curve to cross a road at right angles; and to make the requisite bend to enable it to do so would have taken the railway considerably out of its direct course. To overcome this difficulty "skew bridges" were designed, that crossed roads and canals in slanting directions. Iron girder bridges were also constructed, and thus the railway trains were carried across roads and narrow rivers at any required inclination, supported on flat beams of iron. Suspension bridges were found to be unfitted, on account of their oscillation, for the passage of locomotive engines; therefore, when it became necessary to carry railways across arms of the sea, or wide navigable rivers, at heights sufficient to allow the largest ships to pass underneath, neither girder bridges nor suspension bridges were suited for the purpose. Then arose the necessity of contriving some form of bridge of extensive span that would be sufficiently strong and rigid for railway trains to pass over them in safety.

The Britannia Bridge, across the Menai Straits, was a triumphant response to the call for a new kind of suspended roadway adapted to the requirements of railways. The tubular principle of construction, designed by Mr. Robert Stephenson, was practically tested by Mr. Fairbairn; and the result of numerous experiments on the strength of iron, in different forms and combinations, established the soundness of that principle. The rigidity and strength of the Britannia Bridge depend on cellular cavities at the top and bottom, which, acting as so many tubes, give stability to the riveted plates of iron, and enable the bridge to bear the immense pressure and vibration of a heavy railway train without deflecting more than half an inch.

It was Mr. Stephenson's original intention to make a circular or oval tube, suspended by chains, for the trains to run through; but Mr. Fairbairn's experiments proved that a rectangular shape is stronger, provided the top and bottom, which bear the greatest part of the strain, are made rigid, either by additional plates of iron, or by tubes. The notion of a circular tube was, therefore, abandoned, and the rectangular form, with cells at the top and bottom, was adopted; first for the railway bridge at Conway, and afterwards for the much greater work across the Menai Straits.

It has been stated by Mr. Stephenson, that the idea of forming a tubular bridge was suggested by experience gained in constructing the railway bridge at Ware, which consisted of a wrought-iron cellular platform; but a more exact representation of the principle on which the Britannia Bridge is constructed had been long previously seen across the Rhine, at Schauffhausen, where a rectangular tube, or hollow girder, made of wood, was erected in 1757. That bridge, though of different material, was in its principle of construction similar to the iron tubular bridges at Conway and at the Menai Straits. Another similar bridge, carried over the river Limmat, at Wettingen, constructed in 1778, had a span of 390 feet; and that, as well as the former, was raised to its position in one piece, by means of powerful screw-jaws. These curious and interesting structures, which may be considered the forerunners of the gigantic iron Tubular Bridges of the present day, were burnt by the French in 1799.

In constructing the Britannia Bridge, Mr. Stephenson took advantage of a rock midway from shore to shore, whereon to erect the central pier. Two other piers, at a distance, on each side, of 460 feet, were built without much difficulty in shallower water, and between these and the masonry on each side was a distance of 230 feet. There are eight rectangular tubes resting on those piers, to form two lines of railway, each tube being 28 feet high and 14 feet wide, exclusive of the cellular cavities at the top and bottom. These cavities are rectangular, and extend from one end of the bridge to the other, and may be regarded as long tubes. There are eight of them at the top, each 1 foot 9 inches square, and there are six at the bottom, the latter being 2 feet 4 inches wide, and the same depth as those at the top. Sound is conveyed through these cavities as readily as through speaking tubes, and conversation can be thus easily carried on across the Straits.

The height of the central pier of the Britannia Bridge, from the foundation to the top, is 230 feet; and the height of the roadway above high water mark is 104 feet. The length of the large tubes, through which the railway carriages pass, on each side of the central pier, is 460 feet: and the total length from shore to shore, 1,531 feet. The tubes are connected together at the piers to give the bridge additional strength, and they are composed altogether of 186,000 separate pieces of iron, which were pierced with seven millions of holes, and united together by upwards of two millions of rivets. The whole mass of iron employed weighed 10,540 tons.

The Britannia Bridge was commenced in May, 1846, and the first of the main tubes was completed in June, 1849. The work was carried on close to the bridge, on the Anglesea shore; and when the tube was ready to be transported to its place on the piers, which had been prepared to receive it, eight flat-bottomed pontoons were provided to carry it, which, being brought underneath, floated the ponderous mass on the water as they rose with the tide.

The floating and fixing in its place of the tube took place on the 27th of the same month, in view of an immense concourse of spectators. After the preliminary arrangements for letting go had been completed, Mr. Stephenson, and other engineers, got on the tube, with Captain Claxton, R. N., to whom the management of the floating was entrusted. A correspondent of the *Illustrated London News* thus describes the proceeding, and its successful result:—"Captain Claxton was easily distinguished by his speaking trumpet, and there were also men to hold the letters which indicated the different capstans, so that no mistake could occur as to which capstan should be worked; and flags, red, blue, and white, signified what particular movement should be made. About 7.30 p.m. the first perceptible motion, which indicated that the tide was lifting the mass, was observed, and at Mr. Stephenson's desire, the depth of water was ascertained, and the exact time noted. In a few minutes the motion was plainly visible, the tube being fairly moved forward some inches. This moment was one of intense interest, the huge bulk gliding as gently and easily forward as if she had been but a small boat. The spectators seemed spell-bound, for no shouts or exclamations were heard, as all watched anxiously the silent course of the heavily freighted pontoons. The only sounds heard were the shouts of Captain Claxton, as he gave directions to 'let go ropes,' to 'haul in faster,' &c.; and 'broad-side on,' the tube floated majestically in the centre of the stream. I then left my station, and ran to the entrance of the works, where I got into a boat, and bade the men pull out as far as they could into the middle of the Straits. This was no easy task, the tide running strong; but it afforded me several splendid views of the floating mass, and one was especially fine; the tube coming direct on through the stream —the distant hills covered with trees, two or three small vessels and a steamer, its smoke blending well with the scene, forming a capital background; whilst on one side, in long stretching perspective, stood the three unfinished tubes, destined ere long to form, with the one then speeding on its journey, one grand and unique roadway. It was impossible to see this grand and imposing sight, and not to feel its singleness, if we may so speak. Anything so mighty of its kind had never been before: again it would assuredly be; but it was like the first voyage made by the first steam-vessl—something until then unique. At 8.35 the tube was nearing the Anglesea pier, and at this moment the expectation of the spectators was greatly increased, as the tube was so near its destination: and soon all fears were dispelled, as the Anglesea end of the tube passed beyond the pier, and then the Britannia pier end neared its appointed spot, and it was instantly drawn back close to the recess, so as to rest on the bearing intended for it. There was then a pause for a few minutes, while waiting for the tide to turn: and when that took place, the huge bulk floated gently into its place on the Anglesea pier, rested on the bearing there, and was instantly made fast, so that it could not move again. The cheering, till now subdued, was loud and hearty, and some pieces of cannon on the shore gave token, by their loud booming, that the great task of the day was done."

The tube, when in position, was lowered down upon its bearings on the pier by opening valves in the pontoons, which thus sunk sufficiently to ease them of their load.

The work of raising the tube to its position, 100 feet above high water mark, was a much slower operation, and was attended with serious difficulties. Hydraulic presses were used for the purpose, placed at the top of the piers; two smaller ones, which had served to raise the Conway Bridge, being at one end, and a much larger press, made for the occasion, being fixed at the other. The immense tube was lifted by chains fixed to the heads of the presses, and two steam engines, of 40-horse power each, were employed to force the water into the cylinders. The diameter of the ram of the largest hydraulic press was 20 inches, and the pressure upon it was equal to $2\frac{1}{4}$ tons on each circular inch. The tube was raised by successive lifts of 6 feet each, and, as it was lifted, the space was built in with masonry for its ultimate bearing. During the operation of lifting, the bottom of the cylinder of the large hydraulic press burst out, and fell on the top of the tube, in which it made a considerable indentation. Mr. Stephenson had provided against the possibility of such accident, by having blocks of wood, an inch thick, introduced under the tube as it was elevated, and these blocks arrested its fall, or it would otherwise have been dashed to pieces. Even the small fall of an inch did considerable injury. This accident caused some delay, but the other tubes were in the meantime progressing, and the completed bridge was opened for public traffic on the 21st of October, 1850.

The strength of the bridge was tested before passenger trains were allowed to pass through it, by placing in the centre of the longest tubes twenty-eight waggons, loaded with 280 tons of coal, and two locomotives, and by afterwards sending those heavy trains through the bridge at full speed. The deflection of the tubes in the centre amounted to only three-quarters of an inch in each cell; it being rather less when the trains were at full speed than when stationary. The strongest gusts of wind to which the bridge has been exposed have not caused a vibration of more than one inch. The total cost of construction was £601,865; of which sum £3,986 was for experiments, and £158,704 for masonry.

Another Tubular Bridge of rival magnitude to the one across the Menai Straits is now in the course of construction by Mr. Brunel across the Tamar, at Saltash, for the South Devon and Cornwall Railway. As no rock presented itself conveniently halfway across whereon to erect the central pier, Mr. Brunel was obliged to work at a great depth below the surface of the water in making the foundation of the Royal Albert Bridge. In the plan of making the foundation, as well as in the structure of the bridge itself, Mr. Brunel adopted a course altogether original. Instead of attempting to construct a coffer-dam by piles, which would have been almost impracticable at such a depth, and very costly, he caused a large iron tube to be put together, thirty-six feet in diameter, and ninety-six feet long, to reach to the bed of the river. This monster tube was lowered perpendicularly in the middle of the river, and the water being pumped out of it, the men could work at the bottom in safety. In this manner, after much labour, the rock was prepared to receive the blocks of granite, which were laid one on the other, till they rose above the surface of the water. On that granite pedestal a cast-iron pier was raised to a height of 100 feet, the level of the roadway of the rails.

The cast-iron pier consists of four octagon columns, 10 feet in diameter. They stand about 10 feet apart, forming a square, and they are bound together by massive lattice-work of wrought iron, to prevent any lateral movement. Each of these columns weighs 150 tons; and when the full weight of the bridge rests on the foundation of the central pier, the pressure will be equal to 8 tons on the square foot, or double the pressure of the Victoria Tower on its base.

In the structure of the bridge, Mr. Brunel availed himself of the results of the experiments made by Mr. Fairbairn on the strength of iron tubes, but he adopted a very different plan from that of Mr. Stephenson. Instead of constructing a large tube for the trains to pass through, Mr. Brunel made tubular arches, consisting of iron plates curved and riveted together, to serve as rigid supports, from which the roadway is suspended by chains and by connecting iron bars.

The placing of the first of the tubular arches in position between the pier near the shore at Saltash and the central pier, which took place on the 1st of

September, 1857, excited great interest, and at least 50,000 persons were assembled from places far and near to witness the operation. The tube, with the roadway and suspension chains, was floated from the yard where it was put together on four pontoons; and it was thus conveyed, and safely deposited on the piers at a height of 30 feet above high water mark. It was afterwards gradually raised by hydraulic presses to the top, a height of 100 feet. The work of raising it commenced on the 25th of November, and was completed on the 19th of May last.

The following lively description of the Royal Albert Bridge, and its surrounding scenery, extracted from a recent article in the *Times*, gives a very good idea of the magnitude of the structure, by comparison with well-known objects:—"Though, probably, our readers may care little and have heard less about Saltash proper, it is likely henceforth to receive a fair share of general attention, and we can safely say, to those who will journey down to see the bridge, that the viaduct requires indeed to be a fine one to attract their attention from the lovely scenery of the valley of the Tamar, which it crosses. The banks of this noble river narrow in considerably as the stream reaches Saltash, and, hemmed in there to half a mile or so, suddenly widens out into as fine a sheet of water as any of its kind in the kingdom, its distant banks covered with cottages, and fringed with undulating woodlands down to the very edge. Across this narrow part of the channel, where Saltash, in picturesque dirt and disarray, straggles up the banks on one side, and a steep hill, covered with rock and rock-grown underwood, forms the other, the viaduct stretches high in air. The briefest general way of describing it is to say that it consists of nineteen spans or arches, seventeen of which are wider than the widest arches of Westminster Bridge; and two, resting on a single cast-iron pier of four columns in the centre of the river, span the whole stream at one gigantic leap of 910 feet, or a longer distance than the breadth of the Thames at Westminster. The total length of the structure from end to end is 2,240 feet,--very nearly half a mile, and 300 feet longer than the entire stretch of the Britannia Bridge. The greatest width is only 30 feet at basement; its greatest height from foundation to summit no less than 260 feet, or 50 feet higher than the summit of the Monument. The Britannia Bridge, both in size, purpose, and engineering importance, seems to offer the best comparison with that of Saltash, but the similarity between the structures is far from being as great as might be at first supposed. The Britannia tube is smaller, and cost nearly four times the price of the Saltash Viaduct, though the engineers had natural facilities which Mr. Brunel, for his Cornish bridge, certainly had not."

The form of the tubes is an oval, 17 feet in its longest diameter, and 12 feet in its shortest. They are bent into an elliptical curve, with a rise in the middle of twenty-eight feet. With the roadway and suspension chains attached, each tube weighs 1,100 tons. The total weight of wrought iron in the bridge, when completed, will be 2,650 tons; of cast iron,

1,200; of masonry and brickwork there will be about 17,000 cubic yards; and of timber, about 14,000 cubic feet.

The second tube, which is in every respect like the first, was completed on the 30th of June last, and on the 10th of July was successfully placed in position between the central pier and the Devonshire side of the river. The operation of elevating it began on the 9th of August, and it has now reached nearly the level of the first one, the tube being raised six feet in a week.

The engraving on the other side is a view of this wonderful structure in its completed form. Its appearance is far more light and elegant than that of the Britannia Bridge, but it remains to be seen whether it will be equally steady under a gale of wind, and whether any vibration of the suspended roadway will interfere with the rapid motion of the trains. As the South Devon Railway has only one line of rails for the greater portion of its length, but a single roadway is provided on the Royal Albert Bridge.

The progress of railway locomotion has not only given rise to the construction of new kinds of bridges, but it has directed mechanical science to devise better means of applying the strength of materials. On the South Devon and Cornwall Railways are to be seen wooden viaducts, carrying the line over valleys at great heights, constructed with such slender timbers, that, to an inexperienced eye, they seem frightfully frail for the support of heavy railway trains.

We must not omit to notice, among the remarkable

ROYAL ALBERT BRIDGE, OVER THE TAMAR, AT SALTASH.

bridge erections connected with railways, the viaduct across the valley of the Boyne, which passes over the river close to the town of Drogheda, at a height of 95 feet. The central portion of the viaduct is supported on four piers, 90 feet above high water mark, with a span in the centre of 250 feet, and on each side of 125 feet. This elevated portion of the work is approached on the southern side by twelve arches, of 60 feet span each, and on the north by three similar arches. The viaduct is constructed of limestone and iron lattice-work, and is calculated to bear 7,200 tons.

During the erection of this viaduct the railway trains were carried over the valley on a wooden platform, without side railings, supported by scaffold-poles; and the crackling of the timbers, as the carriages passed over it, and the dizzy height at which they were carried through the air, produced a sensation of terror in nervous passengers, that was fully justified by the apparent danger.

SELF-ACTING ENGINES.

The manufacturing progress of this country has depended, in a great degree, on the facility possessed of making machinery of all kinds by the aid of powerful engines worked by steam power. These engines, most of which appear to be self-acting, forge and roll and cut and bore beams of iron, boiler plates, and cylinders of immense size, which it would be impossible to make by hand; and they do the work with a rapidity and mechanical accuracy that would be otherwise unattainable. In the progress of manufacturing invention, the small steam engine first made by manual labour created the power to make other steam engines of large size; and those more powerful engines supplied the means of making still larger shafts and cylinders for engines that were to be employed in the construction of machines of various kinds, to be worked by the power thus accumulated.

The important advantages derived from the invention and application of self-acting machinery, not only by the community at large, but even by the workmen whose labour they for a time superseded, were forcibly stated by Mr. Whitworth, in his opening address at the Institution of Mechanical Engineers, in September, 1856:—"I congratulate you," he observed, "on the success which in our time the mechanical arts have obtained, and the high consideration in which they are held. Inventors are not now persecuted, as formerly, by those who fancied that their inventions and discoveries were prejudicial to the general interest, and calculated to deprive labour of its fair reward. Some of us are old enough to remember the hostility manifested to the working of the power-loom, the self-acting mule, the machinery for shearing woollen cloth, the thrashing machine, and many others. Now the introduction of reaping and mowing machines, and other improved agricultural machinery, is not opposed. Indeed, it must be obvious, to reflecting minds, that the increased luxuries and comforts which all more or less enjoy, are derived from the numerous recent mechanical appliances and the productions of our manufactories. That of our cotton has increased during the last few years in a wonderful degree. In 1824, a gentleman with whom I am acquainted sold on one occasion 100,000 pieces of 74-reed printing cloth at 30s. 6d. per piece of 29 yards long; the same description of cloth he sold last week at 3s. 9d. One of the most striking instances I know of the vast superiority of machinery over simple instruments used by hand, is in the manufacture of lace, when one man, with a machine, does the work of 8,000 lace makers on the cushion. In spinning fine numbers of yarn, a workman in a self-acting mule will do the work of 3,000 hand-spinners with the distaff and spindle.

"Comparatively few persons, perhaps, are aware of the increase of production in our life-time. Thirty years ago, the cost of labour for turning a surface of cast iron, by chipping and filing with the hand, was 12s. per square foot—the same work is now done by the planing machine at a cost for labour of less than one penny per square foot: and this, as you know, is one of the most important operations in mechanics; it is, therefore, well adapted to illustrate what our progress has been. At the same time that this increased production is taking place, the fixed capital of the country is, as a necessary consequence, augmented; for in the case I have mentioned, of chipping and filing by the hand, when the cost of labour was 12s. per foot, the capital required for tools for one workman was only a few shillings; but now, the labour being lowered to a penny per foot, a capital in planing machines for the workman is required which often amounts to £500, and in some cases more."

Notwithstanding the great economy of labour by the self-acting machines now employed for doing all kinds of work, it is gratifying to find that it has not had the effect of throwing men out of employ; for the increased demand, consequent on the facility of production, has more than compensated for the substitution of automaton mechanism for handicraft.

It is extremely interesting to visit a large engineering factory, and to witness the ease with which the masses of crude metal are wrought in various ways, and converted by a number of seemingly self-acting engines into other engines and machines which are, in their turn, to become the agents of the further development of the skill and ingenuity of man. In the new Government factory at Keyham, near Devonport, which we believe to be one of the largest establishments of the kind in the world, most of those powerful engines of the best construction may be seen in operation. The completeness of the arrangements redounds much to the credit of Mr. Trickett, the chief engineer, under whose supervision they were made; and a walk through the factory, which is thrown open to public inspection, will well repay a journey of many miles. A detailed description of all its machinery would fill a volume, but we must now limit ourselves to a bare enumeration of some of the most remarkable features.

Numerous machines of the largest size, placed under the cover of an extensive and lofty roof, are employed in doing everything requisite for the fitting out of the largest steam-ships in the British navy. Shears, put in continuous motion by steam power, are seen moving steadily up and down, and cutting through the thickest boiler plates without the least apparent effort, the chisel-shaped knives that cut the metal moving just the same whether they be dividing the air or shearing iron. Punching engines, in like manner, force holes through iron plates an inch thick. Shaping and planing machines pare off the tough iron as if it were not harder than cheese. Riveting machines of different kinds bind together the plates of monster boilers with marvellous rapidity; whilst machines for boring, for drilling, for forging, and for doing every variety of smaller work, are to be seen in operation in various parts of the factory.

Among the smaller self-acting engines, the forging machine for making bolts attracts attention by the rapidity of its action. It consists of a series of hammers placed side by side, so constructed as to shape small bars of iron into any required form, according to the mould of the swages beneath them, representing miniature anvils. It is interesting to watch how readily the hot iron receives its shape under the action of the hammers, which make about 700 strokes per minute, the work being transferred from one to another to be progressively finished. There is a circular saw that cuts through bars of iron as thick as railway rails, by making upwards of 1,000 revolutions per minute. A rivet-making machine forms the rivet, and shapes the head to the requisite size, with great accuracy and quickness. There are compound drilling machines, in which six drills are acting simultaneously; hydraulic presses, that force parts of machines together, and a great variety of other engines for the saving of time and labour.

Not the least curious of the smaller contrivances is an apparatus which deserves notice as a useful application of magnetism to manufacturing purposes. Several horse-shoe magnets are attached to two endless chains, moving over suitable wheels, and inclined at an angle of 30 degrees. These magnets at the lower end of the chain, dip into a tub containing the mixed brass and iron turnings and filings from the lathes and other tools, and the pieces of iron, being attracted by the magnets, are carried away and brushed off into a box, leaving the brass behind to be remelted.

In one department of the building are immense foundry furnaces, where metals are melted and cast, the blast of the fires being maintained by large rotating fans, kept in action by a powerful steam engine, by which also the other machines are worked. The foundry is most conveniently contrived for casting works of any required size, fixed and travelling cranes being so stationed and arranged as to carry the ladles of liquid metal to any part of the floor.

In another department is the smithy, where the iron to be wrought into shape is heated in forges; and near to the forges stand the Steam-Hammers— those gigantic Cyclops of modern times, that strike blows, compared with the force of which the blows of the fabled Cyclops of antiquity were but as the fall of a feather.

Ranged in a row there are four of these ponderous engines, of various sizes; the largest hammer being so heavy as to require the power of four tons to lift it, and when falling from a height of 6 feet nothing can withstand its crushing blow. Yet the force of this mighty giant is so completely under control, and may be brought to act so gently, as scarcely to crack a nut placed to receive its fall.

The invention of the steam-hammer was the result of necessity. The shaft of a steam engine having to be made larger than usual, no hammer then in action by water power was capable of forging it, and Mr.

James Nasmyth was applied to, to give his aid in contriving the means of removing the difficulty. It was then that the idea of lifting the hammer-block by the direct action of steam occurred to him, and by a succession of extremely ingenious devices, he at length perfected the steam-hammer, which has been pronounced to be one of the most perfect artificial machines, and one of the noblest triumphs of mind over matter that modern English engineers have yet developed.

The accompanying woodcut represents the largest of the four steam-hammers in Keyham factory. The hammer block, *a*, weighing four tons, is guided in its ascent and fall by grooves in two massive uprights, which hold the whole together. The hammer-block is lifted by the piston rod of the steam cylinder above it, which is made of such diameter, that the pressure of the steam on the surface of the piston may considerably overbalance the weight of the hammer-block, and overcome the friction of the connecting mechanism. The cylinder of the largest steam-hammer at Keyham is 18 inches diameter, which gives an area of 254 square inches; and the pressure of the steam generally used being fifty pounds on the square inch, the total steam pressure tending to force the piston up, when the whole of it is brought to bear, is equal to five tons and a half. The force of the blow of the hammer, when falling from its greatest height, is equal to 144 tons.

By the arrangements of levers, screws, and pipes and valves, shown in the engraving, the steam is first admitted under the piston, and thus acts directly in forcing it up, with the heavy hammer-block attached to the piston rod. When the block has been raised to the required height, it strikes against the end of a

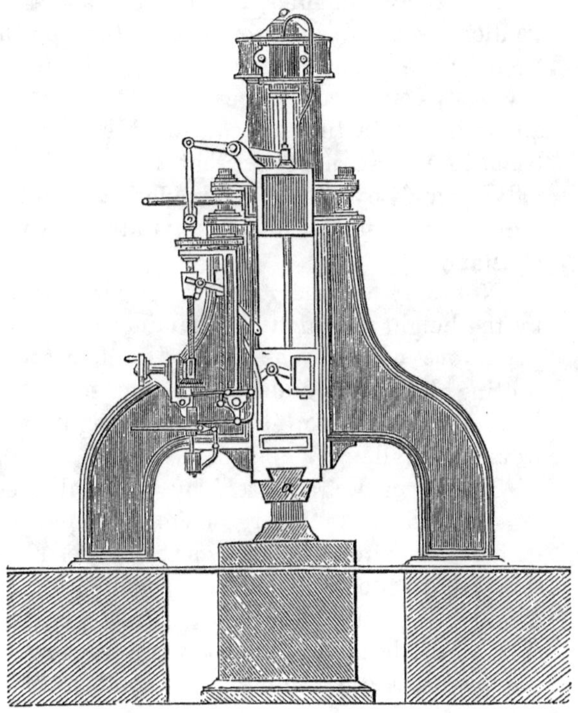

lever, which then shuts off the steam, and allows it to escape; whereupon the hammer falls with its full force vertically on the anvil. The end of the lever which turns off the steam may be adjusted at any height, according to the required force of the blow, so that the hammer may fall from a height of six feet, or be merely raised a few inches.

The steam hammer, in the early stages of its invention, required an attendant to turn on the steam again at the end of each stroke, but Mr. Nasmyth ingeniously contrived the means of rendering the engine altogether self-acting, by causing the force of the collision to release a spring that holds down the slide-valve; and by this contrivance a continued and regular succession of blows is maintained without any assistance.

Not only can the force of the blow be regulated by the height to which the hammer is lifted, but the ponderous mass may be arrested in its descent by admitting the steam under the piston, so that a skilful manipulator can stop it within the eighth of an inch from the anvil.

The Steam Engine itself, by which all the self-acting mechanisms of a large factory are put in motion, is, perhaps, after all, the most wonderful of inventions; but it does not strictly come within our province, for Watt had perfected his great work before the close of the last century. It was, however, not much used, excepting for mining purposes, until after the commencement of the present; and the inventor himself had but a faint idea of the value and vast importance of the motive power he had placed at the command of man. So little, indeed, was the value of steam power appreciated in the early years of its application, that no notice is taken of the steam engine in Beckmann's History of Inventions, though Watt had completed his condensing engines several years before that work was published; and Newcomen's steam engine had been at work at least sixty years.

The history of the steam engine affords a striking example of the gradual development of an invention from vague and chimerical notions, into an accomplished fact of astonishing magnitude. As in the electric telegraph the dreams of the alchemist are fully realized by the applications of scientific discovery, so in the wonder-working powers of the steam engine one of the visionary schemes sketched in the "Century of Inventions" is practically extended far beyond the conceptions of its fanciful projector. How little could Beckmann have supposed that an invention, which he considered too insignificant to be mentioned, would, in the course of fifty years, have revolutionized the world! It may possibly be the same, before this century is closed, with inventions that are now neglected or despised.

The record in the preceding pages of some of the most remarkable applications of science during the present century, exhibits an amount of intelligence, of skill, and of power that seems, when viewed in its completed form, to be superhuman. It is only by tracing each invention to its source, and by noting the step by step advances by which it has arrived at its present state, that we can bring ourselves to believe that the great development of power and the display of ingenuity we witness, can have been accomplished by ordinary men. This feeling of admiration, at the results of human industry and inventive genius, was strongly excited on passing through the wonderful collection of the works of all nations in the Great Exhibition of 1851. After walking through the long avenues, crowded with the most highly finished manufactured goods, and with works of art, adapted to every purpose and capable of gratifying every luxurious taste of highly civilized life, we beheld, in another part of the building, the self-acting machines by which many of those productions had been manufactured. We saw various mechanisms, moving without hands to guide them, producing the most elaborate works; massive steam engines,—the representatives of man's power,—and exquisite contrivances, displaying his ingenuity and perseverance; and we felt inclined to exalt the attributes of humanity, and to think that nothing could surpass the productions there displayed. But as if to repress such vainglorious thoughts, there stood in the transept of the building, surrounded by and contrasting with the handiworks of man, one of the simplest productions of Nature. Every single leaf on the spreading branches of that magnificent tree exhibited in its structure, in its self-supporting and self-acting mechanism, and in the adaptation of surrounding circumstances for its maintenance, an amount of intelligent design and contrivance and power, with which there was nothing to compare. After examining the intricate ramifications of arteries and veins for spreading the sap throughout the leaf, and the innumerable pores for inhaling and exuding the gases and moisture necessary for its continued existence ; after carrying the mind beyond the beautiful structure itself, to consider the provisions of heat and moisture and air, without which all that mechanism would have been useless ; and having reflected on the presence of the mysterious principle which actuated the whole arrangement of fibres, and gave life to the crude elements of matter,—we could not fail to be impressed with the insignificance of the most elaborate productions of man, when compared with the smallest work of the Omnipotent Creator.

THE END.

Chemical Chart:

BY E. L. YOUMANS.

On Rollers, 5 feet by 6 in size. New Edition. Price $5.

This popular work accomplishes for the first time, for Chemistry, what maps and charts have for geography, geology, and astronomy, by presenting a new and valuable mode of illustration. Its plan is to represent chemical composition to the eye by colored diagrams, so that numerous facts of proportion, structure, and relation, which are the most difficult in the science, are presented to the mind through the medium of the eye, and may thus be easily acquired and long retained. The want of such a chart has long been felt by the thoughtful teacher, and no other scientific publication that has ever emanated from the American press has met with the universal favor that has been accorded to this Chart. In the language of a distinguished chemist, "Its appearance marks an era in the progress of the popularization of Chemistry."

It illustrates the nature of elements, compounds, affinity, definite and multiple proportions, acids, bases, salts, the salt-radical theory, double decomposition, deoxidation, combustion and illumination, isomerism, compound radicals, and the composition of the proximate principles of food. It covers the whole field of Agricultural Chemistry, and is invaluable as an aid to public lecturers, to teachers in class-room recitation, and for reference in the family. The mode of using it is explained in the class-book.

From the late HORACE MANN, *President of Antioch College.*

I think Mr. Youmans is entitled to great credit for the preparation of his Chart, because its use will not only facilitate acquisition, but, what is of far greater importance, will increase the exactness and precision of the student's elementary ideas.

From DR. JOHN W. DRAPER, *Professor of Chemistry in the University of N. Y.*

Mr. Youmans' Chart seems to me well adapted to communicate to beginners a knowledge of the definite combinations of chemical substances, and as a preliminary to the use of symbols, to aid them very much in the recollection of the examples it contains. It deserves to be introduced into the schools.

Fron JAMES B. ROGERS, *Professor of Chemistry in the University of Pennsylvania.*

We cordially subscribe to the opinion of Professor Draper concerning the value to beginners of Mr. Youmans' Chemical Chart.

JOHN TORREY,
Professor of Chemistry in the College of Physicians & Surgeons, N. Y.

WM. H. ELLET,
Late Professor of Chemistry in Columbia College, S. C.

JAMES B. ROGERS,
Professor of Chemistry in the University of Pennsylvania.

From BENJAMIN SILLIMAN, LL. D., *Professor of Chemistry in Yale College.*

I have hastily examined Mr. Youmans' New Chemical Diagrams or Chart of chemical combinations by the union of the elements in atomic proportions. The design appears to be an excellent one.

A History of Philosophy:

AN EPITOME.

BY DR. ALBERT SCHWEGLER.

TRANSLATED FROM THE ORIGINAL GERMAN, BY JULIUS H. SEELYE.

12mo. 365 pages. Price $1 25.

This translation is designed to supply a want long felt by both teachers and students in our American colleges. We have valuable histories of Philosophy in English, but no *manual* on this subject so clear, concise, and comprehensive as the one now presented. Schwegler's work bears the marks of great learning, and is evidently written by one who has not only studied the original sources for such a history, but has thought out for himself the systems of which he treats. He has thus seized upon the real germ of each system, and traced its process of development with great clearness and accuracy. The whole history of speculation, from Thales to the present time, is presented in its consecutive order. This rich and important field of study, hitherto so greatly neglected, will, it is hoped, receive a new impulse among American students through Mr. Seelye's translation. It is a book, moreover, invaluable for reference, and should be in the possession of every public and private library.

From L. P. Hickok, *Vice-President of Union College.*

"I have had opportunity to hear a large part of Rev. Mr. Seelye's translation of Schwegler's History of Philosophy read from manuscript, and I do not hesitate to say that it is a faithful, clear, and remarkably precise English rendering of this invaluable Epitome of the History of Philosophy. It is exceedingly desirable that it should be given to American students of philosophy in the English language, and I have no expectation of its more favorable and successful accomplishment than in this present attempt. I should immediately introduce it as as a text-book in the graduate's department under my own instruction, if it be favorably published, and cannot doubt that other teachers will rejoice to avail themselves of the like assistance from it."

From Henry B. Smith, *Professor of Christian Theology, Union Theological Seminary, N. Y.*

"It will well reward diligent study, and is one of the best works for a text-book in our colleges upon this neglected branch of scientific investigation."

From N. Porter, *Professor of Intellectual Philosophy in Yale College.*

"It is the only book translated from the German which professes to give an account of the recent German systems which seems adapted to give any intelligible information on the subject to a novice."

From Geo. P. Fisher, *Professor of Divinity in Yale College.*

"It is really the best Epitome of the History of Philosophy now accessible to the English student."

From Joseph Haven, *Professor of Mental Philosophy in Amherst College.*

"As a manual and brief summary of the whole range of speculative inquiry, I know of no work which strikes me more favorably."

Moral Philosophy.

Elements of Moral Philosophy:

ANALYTICAL, SYNTHETICAL, AND PRACTICAL.

BY HUBBARD WINSLOW.

12mo. 480 pages. Price $1 25.

This work is an original and thorough examination of the fundamental laws of Moral Science, and of their relations to Christianity and to practical life. It has already taken a firm stand among our highest works of literature and science. From the numerous commendations of it by our most learned and competent men, we have room for only the following brief extracts:

From the REV. THOMAS H. SKINNER, D. D., *of the Union Theol. Sem., N. Y.*

"It is a work of uncommon merit, on a subject very difficult to be treated well. His analysis is complete. He has shunned no question which his purpose required him to answer, and he has met no adversary which he has not overcome."

From REV. L. P. HICKOK, *Vice-President of Union College.*

"I deem the book well adapted to the ends proposed in the preface. The style is clear, the thoughts perspicuous. I think it calculated to do good, to promote the truth, to diffuse light, and impart instruction to the community, in a department of study of the deepest interest to mankind."

From REV. JAMES WALKER, D. D., *President of Harvard University.*

"Having carefully examined the more critical parts, to which my attention has been especially directed, I am free to express my conviction of the great clearness, discrimination, and accuracy of the work, and of its admirable adaptation to its object."

From REV. RAY PALMER, D. D., *of Albany.*

"I have examined this work with great pleasure, and do not hesitate to say that in my judgment it is greatly superior to any treatise I have seen, in all the essential requisites of a good text-book."

From PROF. ROUSSEAU D. HITCHCOCK, D. D., *of Union Theol. Sem., N. Y.*

"The task of mediating between science and the popular mind, is one that requires a peculiar gift of perspicuity, both in thought and style; and this, I think, the author possesses in an eminent degree. I am pleased with its comprehensiveness, its plainness, and its fidelity to the Christian stand-point."

From PROF. HENRY B. SMITH, D. D., *of the Union Theol. Sem., N. Y.*

"It commends itself by its clear arrangement of the topics, its perspicuity of language, and its constant practical bearings. I am particularly pleased with its views of conscience. Its frequent and pertinent illustrations, and the Scriptural character of its explanations of the particular duties, will make the work both attractive and valuable as a text-book, in imparting instruction upon this vital part of philosophy."

From W. D. WILSON, D. D., *Professor of Intellectual and Moral Philosophy in Hobart Free College.*

"I have examined the work with care, and have adopted it as a text-book in the study of Moral Science. I consider it not only sound in doctrine, but clear and systematic in method, and withal pervaded with a prevailing healthy tone of sentiment, which cannot fail to leave behind, in addition to the truths it inculcates, an impression in favor of those truths. I esteem this one of the greatest merits of the book. In this respect it has no equal, so far as I know; and I do not hesitate to speak of it as being preferable to any other work yet published, for use in all institutions where Moral Philosophy forms a department in the course of instruction."

www.ingramcontent.com/pod-product-compliance
Lightning Source LLC
LaVergne TN
LVHW020220110826
845151LV00003B/778

* 9 7 8 1 4 2 5 5 3 2 1 3 0 *